Deepen Your Mind

序言

近年來，Java 增加了函數式程式設計的特性，如類型推斷、Lambda 運算式、Stream 流等，後端開發逐漸採用了一些函數式程式語言，如 Scala、Kotlin。函數式程式設計以其簡潔性、不變性、空指標處理人性化等特點深受後端開發人員的青睞。

筆者之前使用 Scala 做過後端開發，感慨 Scala 的門檻較高，入門困難，而且生態資源相對匱乏，開發過程比較痛苦。後來，使用 Java 進行後端開發，依靠 Spring Boot 強大的生態，可以方便地使用訊息佇列、資料庫、快取、大數據相關中介軟體。但是 Java 比較笨重，容錯的程式、空指標異常、執行緒安全等問題經常困擾著筆者。隨著 Kotlin 在行動端開發的普及，它也逐步走入後端開發者的視野。Kotlin 是 JVM 系統的語言，和 Java 具有良好的互通性，上手較容易，且可以使用 Java 強大的生態，其還具有函數式程式設計的優點。另外，Spring Initializr 提供了對 Java、Kotlin 語言的支援。

Kotlin 是 JetBrains 公司開發的，目前流行的 IntelliJ IDEA 軟體也是該公司開發的。IDEA 對 Kotlin 支援較好，可以將 Java 程式轉為 Kotlin 程式。IDEA 還支援 Java、Kotlin 混合程式設計，歷史程式使用 Java 撰寫，新的程式可以嘗試使用 Kotlin 撰寫。

基於以上考慮，筆者開始研究使用 Kotlin、Spring Boot 做後端開發，獲得了不錯的效果。市面上介紹使用 Kotlin 進行後端開發的圖書比較少，筆者在大量實作的基礎上，萌生了寫一本書的想法，希望和更多的 Java 開發人員分享 Kotlin 在後端開發中的實作經驗。

本書共 10 章，第 1 章介紹如何架設 Kotlin 的開發環境，第 2 章介紹函數式程式設計，第 3 章簡單介紹 Kotlin 的語法，第 4 章介紹 Kotlin 在常用中介軟體中的應用，第 5 章介紹 Kotlin 如何應用於微服務註冊中心，第 6 章介紹 Kotlin 如何應用於微服務設定中心，第 7 章介紹 Kotlin 如何應用於微服務閘道，第 8 章介紹 Kotlin 如何應用於 Spring Cloud Alibaba，第 9 章介紹 Kotlin 整合服務監控和服務鏈路監控的相關知識，第 10 章介紹如何用 Kotlin 撰寫部落格應用。本書提供了大量的實例，相關原始程式可以從 GitHub 下載運行。

袁康

目錄

01 架設 Kotlin 開發環境

1.1　Kotlin 簡介 1-1

1.2　在 Windows 環境中架設 Kotlin 開發環境 1-6

1.3　在 Ubuntu 環境中架設 Kotlin 開發環境 1-11

1.4　在 macOS 環境中架設 Kotlin 開發環境 1-15

1.5　第一個 Kotlin 程式 1-16

1.6　小結 1-17

02 函數式程式設計介紹

2.1　初識函數式程式設計 2-1

2.2　函數式程式設計的特點 2-4

2.3　Scala、Kotlin、Java 的比較 2-6

2.4　小結 2-8

03 Kotlin 的語法

3.1　基礎語法 3-1

　　3.1.1　基底資料型態 3-1

　　3.1.2　套件名稱和參考 3-6

　　3.1.3　流程控制 3-7

　　3.1.4　傳回和跳躍 3-9

3.2　類別 3-10

　　3.2.1　類別、屬性、介面 3-10

　　3.2.2　特殊類別 3-15

　　3.2.3　泛型 3-17

　　3.2.4　委派 3-19

3.3　函數和 Lambda 運算式 3-21

3.3.1 函數 3-21

3.3.2 Lambda 運算式 3-24

3.4 集合 3-27

3.4.1 集合概述 3-27

3.4.2 集合操作 3-35

3.4.3 List、Set、Map 相關操作 3-43

3.5 程式碼協同 3-45

3.5.1 程式碼協同基礎 3-45

3.5.2 程式碼協同進階 3-50

3.6 小結 3-54

04 Kotlin 在常用中介軟體中的應用

4.1 Kotlin 整合 Spring Boot 4-1

4.1.1 Spring Boot 介紹 4-1

4.1.2 用 Kotlin 開發一個 Spring Boot 專案 4-3

4.2 Kotlin 整合 Redis 4-7

4.2.1 Redis 介紹 4-7

4.2.2 使用 Kotlin 操作 Redis 4-9

4.3 Kotlin 整合 JPA、QueryDSL 4-17

4.3.1 JPA、QueryDSL 介紹 4-17

4.3.2 使用 Kotlin 操作 JPA、QueryDSL 4-19

4.4 Kotlin 整合 MongoDB 4-28

4.4.1 MongoDB 介紹 4-28

4.4.2 使用 Kotlin 操作 MongoDB 4-30

4.5 Kotlin 整合 Spring Security 4-36

4.5.1 Spring Security 介紹 4-37

4.5.2 使用 Kotlin 操作 Spring Security 4-38

4.6 Kotlin 整合 RocketMQ 4-46

    4.6.1　RocketMQ 介紹 4-46

    4.6.2　使用 Kotlin 操作 RocketMQ 4-48

4.7　Kotlin 整合 Elasticsearch 4-54

    4.7.1　Elasticsearch 介紹 4-54

    4.7.2　使用 Kotlin 操作 Elasticsearch 4-56

4.8　Kotlin 整合 Swagger 4-62

    4.8.1　Swagger 介紹 4-62

    4.8.2　使用 Kotlin 操作 Swagger 4-64

4.9　小結 4-70

05 Kotlin 應用於微服務註冊中心

5.1　Eureka 5-1

    5.1.1　Eureka 介紹 5-1

    5.1.2　Kotlin 整合 Eureka 服務註冊 5-3

    5.1.3　一個 Eureka 服務提供方 5-6

    5.1.4　Kotlin 整合 OpenFeign 服務呼叫 5-10

    5.1.5　Kotlin 整合 Ribbon 服務呼叫 5-14

5.2　Consul 5-19

    5.2.1　Consul 介紹 5-19

    5.2.2　Kotlin 整合 Consul 服務註冊 5-21

    5.2.3　Kotlin 整合 OpenFeign 和 Ribbon 服務呼叫 5-25

5.3　Zookeeper 5-30

    5.3.1　Zookeeper 介紹 5-30

    5.3.2　Kotlin 整合 Zookeeper 服務註冊 5-32

    5.3.3　Kotlin 整合 OpenFeign 和 Ribbon 服務呼叫 5-36

5.4　Nacos 5-41

    5.4.1　Nacos 介紹 5-41

    5.4.2　Kotlin 整合 Nacos 服務註冊 5-42

    5.4.3　Kotlin 整合 OpenFeign 和 Ribbon 服務呼叫 5-46

5.5　小結 5-50

06 Kotlin 應用於微服務設定中心

6.1　Spring Cloud Config..6-1

　　6.1.1　Spring Cloud Config 介紹...6-1

　　6.1.2　Kotlin 整合 Spring Cloud Config...............................6-3

6.2　Apollo 設定中心..6-10

　　6.2.1　Apollo 介紹...6-11

　　6.2.2　Kotlin 整合 Apollo...6-12

6.3　Nacos 設定中心..6-17

6.4　Consul 設定中心...6-23

6.5　小結..6-28

07 Kotlin 應用於微服務閘道

7.1　Kotlin 整合 Zuul..7-1

　　7.1.1　Zuul 介紹..7-1

　　7.1.2　Kotlin 整合 Zuul...7-3

7.2　Kotlin 整合 Spring Cloud Gateway.....................................7-14

　　7.2.1　Spring Cloud Gateway 介紹.....................................7-15

　　7.2.2　Kotlin 整合 Spring Cloud Gateway...........................7-16

7.3　小結..7-27

08 Kotlin 應用於 Spring Cloud Alibaba

8.1　服務限流降級...8-3

　　8.1.1　Sentinel 介紹..8-3

　　8.1.2　Kotlin 整合 Sentinel...8-5

8.2　訊息驅動...8-12

　　8.2.1　訊息驅動介紹...8-12

　　8.2.2　Kotlin 整合 RocketMQ 實現訊息驅動.......................8-14

8.3　阿里物件雲端儲存..8-22

　　8.3.1　阿里物件雲端儲存介紹..8-22

    8.3.2　Kotlin 整合阿里物件雲端儲存 8-24

8.4　分散式任務排程 8-30

    8.4.1　SchedulerX 介紹 8-31

    8.4.2　Kotlin 整合 SchedulerX 8-32

8.5　分散式交易 8-36

    8.5.1　分散式交易介紹 8-36

    8.5.2　Kotlin 整合 Seata 8-39

8.6　Spring Cloud Dubbo 8-55

    8.6.1　Dubbo 介紹 8-55

    8.6.2　Kotlin 整合 Spring Cloud Dubbo 8-57

8.7　小結 8-66

09 Kotlin 整合服務監控和服務鏈路監控

9.1　Prometheus、Grafana 介紹 9-1

9.2　Kotlin 整合 Prometheus、Grafana 9-4

9.3　Kotlin 整合 Zipkin 9-10

9.4　Kotlin 整合 SkyWalking 9-21

9.5　小結 9-32

10 基於 Kotlin 和 Spring Boot 架設部落格

10.1　初始化 Maven 專案 10-1

10.2　系統架構 10-7

10.3　定義實體 10-8

10.4　資料庫設計 10-18

10.5　Repository 層的設計 10-19

10.6　Service 層的設計 10-42

10.7　Controller 層的設計 10-50

10.8　部署到騰訊雲 10-68

10.9　小結 10-69

Chapter 01

架設 Kotlin 開發環境

本章主要介紹如何在 Windows、Linux、macOS 平台架設 Kotlin 開發環境，包含安裝 JDK、IDEA。本章還將簡單介紹 Kotlin 語言的特性，Kotlin 是一種執行在 Java 平台的函數式程式語言。在本章最後將使用 IDEA 撰寫一個 Kotlin 範例程式。

1.1 Kotlin 簡介

Kotlin 是一種執行在 Java 平台的函數式程式語言，由 JetBrains 公司開發。Kotlin 支援多個平台，包含行動端、服務端及瀏覽器端。Kotlin 歷經了 1.1、1.2 版本，目前最新的版本是 1.3。

Kotlin 融合了物件導向和函數式程式設計，其簡潔、安全、優雅，可以和 Java 完全互動，並可使用 Java 撰寫的協力廠商 Jar 套件。下面我們以一個小實例解釋 Kotlin 是什麼。這個實例定義了一個 Animal 類別，然後建立了一些 Animal 類別的物件，要找出其中 age 最大的 Animal 物件並將其列印出來。在這個實例中，可以看到 Kotlin 的許多特性。程式如下：

```
1.   //1 資料類別
2.   data class Animal(val name: String,
3.                     //2 可為空的類型 (Int?)，宣告變數的預設值
4.                     val age: Int? = null)
5.
6.   //3 頂層函數
7.   fun main(args: Array<String>) {
8.       //4 命名宣告
9.       val animals = listOf(Animal("Dog"), Animal("Cat", 3))
10.
11.      //5 Lambda 運算式
12.      val oldest = animals.maxBy { it.age ?: 0 }
13.      //6 字串範本
14.      println("The Oldest is $oldest")
15.
16.      //7 自動產生 toString 方法
17.      //The Oldest is Animal(name=Cat, age=3)
18. }
```

這裡宣告了一個 Animal 物件，它帶有兩個屬性 name 和 age。age 屬性的
預設值為 null。當建立一個 name=Dog 的 Animal 物件時，沒有設定 age
屬性，預設值是 null。之後，使用 maxBy 方法尋找年齡最大的 Animal 物
件，使用 it 作為預設的參數名稱。如果 age 是 null 的話，設定預設值為
0。由於沒有指定 Dog 的年齡，因此使用預設值 0。最後，Cat 是年齡最大
的 Animal 物件。

透過這個實例，我們對 Kotlin 有了一個初步的印象，下面介紹 Kotlin 的幾
個特性。

❏ 靜態類型

Kotlin 是一種靜態類型程式語言，編譯器可以驗證程式中的變數、運算式
類型。而以 JVM 為基礎的動態類型程式語言，如 Groovy，可以在執行時
期解析方法和欄位參考。另外，Kotlin 不需要在程式中指定每一個變數的

類型，在許多場景中，能夠根據上下文自動推斷變數類型。目前大多數公司仍然在使用 Java 7 或 Java 8，這些 Java 版本沒有這個特性。下面是一個簡單的實例：

```
Val x = 1
```

其中宣告了一個變數，由於它以一個整數初始化，因此 Kotlin 會自動推斷這個變數的類型為 Int。靜態類型有以下好處。

- 效能：由於不需要在執行時期判斷需要呼叫哪個方法，因此方法呼叫速度快。
- 可讀性：由於編譯器驗證了程式的正確性，因此在執行時期發生當機的可能性較小。
- 可維護性：由於能看到程式呼叫了什麼類型的物件，所以可以更容易地了解程式。
- 工具支援：有豐富的 IDE 支援程式設計。

Kotlin 支援可為空的類型（nullable type），可以在編譯時檢查可能的空指標異常，使得程式更加可靠。此外，Kotlin 還支援函數類型（functional type）。

❏ 支援函數式程式設計和物件導向程式設計

函數式程式設計的關鍵概念如下：

- 函數是一等公民。可以把函數看作一個值，把函數儲存在變數中作為一個參數傳遞或傳回。
- 不變性。使用不可變的物件，一旦物件被建立，它的狀態不可更改。
- 沒有副作用。指定相同的輸入將傳回相同的結果。它不會修改其他物件的狀態或和外界進行互動。

用函數撰寫程式簡潔、優雅。將函數作為一個值可帶來更強大的抽象力，避免程式容錯。函數式程式設計是執行緒安全的。如果使用不可修改的資料結構和純函數，可以確保不會出現不安全的修改，也不需要設計複雜的同步方案。Kotlin 有豐富的特性支援函數式程式設計：

- 允許函數接收其他函數作為參數或傳回其他函數。
- 使用 Lambda 運算式，使用最小的範本分發程式區塊。
- 為建立不可變物件提供了精簡的語法。
- 標準函數庫為以函數式風格使用物件和集合提供了豐富的 API。

❏ 實用性

Kotlin 以多年的大規模系統設計的企業經驗為基礎，參考了許多軟體開發者遇到的案例，因此具有極強的實用性。JetBrains 公司和社區的開發者已經使用 Kotlin 早期版本許多年，他們的各種回饋已經融合到發行版本中。Kotlin 不強迫你使用任何特定的程式設計風格或範式。當你使用這門語言時，可以使用自己在進行 Java 開發時熟悉的風格。IntelliJ IDEA 提供了對 Kotlin 強大的支援，可自動將 Java 轉為 Kotlin，可以最佳化程式區塊，並提供了程式填充功能。

❏ 精簡

程式越精簡，人了解得越快，維護越方便。使用 Kotlin，可以簡便地定義資料類別，省略了大量的 getter、setter 和將建構元參數設定值給欄位的邏輯。Kotlin 擁有豐富的標準函數庫，這些函數庫讓你可以透過呼叫函數庫函數來代替那些冗長的、重複的程式片段。Kotlin 對 Lambda 運算式的支援使得將少部分程式傳遞到函數庫函數變得十分容易。這讓你可以將所有通用的部分封裝到函數程式庫，程式中僅保留業務相關的部分。

❏ 安全

使用 Kotlin 可以透過花費較小的代價即達到一個比 Java 更高級別的安全水平。Kotlin 努力從程式中移除空指標異常，其類型系統追蹤可能為空的值，並且禁止執行時期導致空指標異常的操作。它需要的額外成本是最小的：只需要一個單獨的字元，在尾端加一個問號，即標記為一個可能為空的類型。

```
1.  val s: String? =    null          // 1   可能為空
2.  val s2: String  =   ""            // 2   一定不為空
```

另外，Kotlin 提供了許多便捷的方式來處理可能為空的資料，這在避免應用程式當機方面提供了相當大的幫助。

Kotlin 有助避免類型強轉異常。使用 Java 程式設計，開發者經常會遺漏類型檢查，而 Kotlin 將類型檢查和轉換合併為一個單獨的操作，一旦檢查了類型，就可以應用該類型的成員而無須額外進行類型轉換。

```
1.  if(value is String)                // 1   類型檢查
2.      println(value.toUpperCase())   // 2   呼叫該類型的方法
```

❏ 互通性

Kotlin 可以使用已有的 Java 函數庫，呼叫 Java 方法，擴充 Java 類別，實現 Java 介面。同時，也可以在 Java 程式中呼叫 Kotlin 程式。Kotlin 的類別和方法能夠像正常的 Java 類別和方法那樣被呼叫，而且我們還可以在專案的任何地方混合使用 Java 和 Kotlin。

Kotlin 盡可能地使用現有的 Java 函數庫。舉例來說，Kotlin 並沒有自己的集合函數庫，它依賴 Java 標準函數庫的類別，透過額外的函數來擴充它們，這樣就可以更加方便地使用 Kotlin 了。

Kotlin 工具還為多語言專案提供了全面的支援。它能夠編譯任意一個混合 Java 和 Kotlin 的原始檔案，不論它們之間是如何相互依賴的。這個 IDE 特性對其他語言也是有效的。它將允許你做以下事情：

■ 在 Java 和 Kotlin 原始檔案中自由切換。

- 偵錯混合語言專案並在用不同語言撰寫的程式中進行單步追蹤。
- 使用 Kotlin 重構和正確地升級你的 Java 函數。

1.2 在 Windows 環境中架設 Kotlin 開發環境

架設 Kotlin 開發環境需要安裝 Java、IntelliJ IDEA 工具。本書使用 Java 1.8 和 IntelliJ IDEA 2019.2.4 社區版本。IntelliJ IDEA 附帶 Kotlin 外掛程式（Kotlin 的版本是 1.3.60）。

要在 Oracle 官網 (https://www.oracle.com/java/technologies/javase-downloads. html)下載 Java SE Development Kit 8u231，只需點擊下載 jdk-8u231-windows-x64.exe 檔案，如圖 1.1 所示。

Java SE Development Kit 8u231

You must accept the Oracle Technology Network License Agreement for Oracle Java SE to download this software.
Thank you for accepting the Oracle Technology Network License Agreement for Oracle Java SE; you may now download this software.

Product / File Description	File Size	Download
Linux ARM 32 Hard Float ABI	72.9 MB	jdk-8u231-linux-arm32-vfp-hflt.tar.gz
Linux ARM 64 Hard Float ABI	69.8 MB	jdk-8u231-linux-arm64-vfp-hflt.tar.gz
Linux x86	170.93 MB	jdk-8u231-linux-i586.rpm
Linux x86	185.75 MB	jdk-8u231-linux-i586.tar.gz
Linux x64	170.32 MB	jdk-8u231-linux-x64.rpm
Linux x64	185.16 MB	jdk-8u231-linux-x64.tar.gz
Mac OS X x64	253.4 MB	jdk-8u231-macosx-x64.dmg
Solaris SPARC 64-bit (SVR4 package)	132.98 MB	jdk-8u231-solaris-sparcv9.tar.Z
Solaris SPARC 64-bit	94.16 MB	jdk-8u231-solaris-sparcv9.tar.gz
Solaris x64 (SVR4 package)	133.73 MB	jdk-8u231-solaris-x64.tar.Z
Solaris x64	91.96 MB	jdk-8u231-solaris-x64.tar.gz
Windows x86	200.22 MB	jdk-8u231-windows-i586.exe
Windows x64	210.18 MB	jdk-8u231-windows-x64.exe

圖 1.1　Java SE Development Kit 8u231 各平台安裝套件

根據提示，一步步安裝 JDK。選擇 JDK 的安裝位置，如圖 1.2 所示。

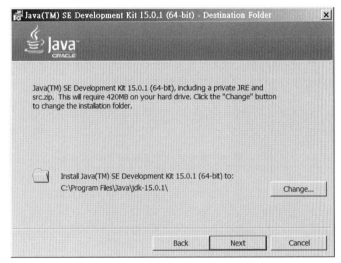

圖 1.2　選擇 JDK 的安裝位置

執行安裝過程，待 JDK 安裝完成後會安裝 JRE，提示選擇 JRE 的安裝位置，如圖 1.3 所示。

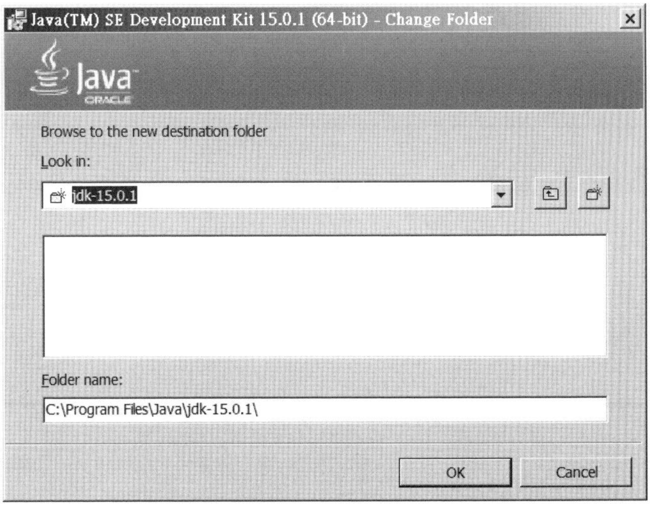

圖 1.3　安裝過程

看到圖 1.4，表示 Java 安裝好了。

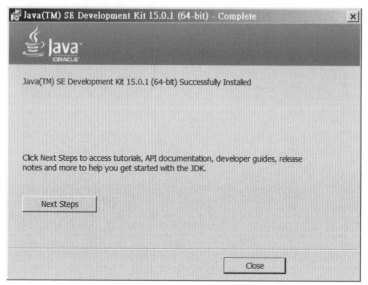

圖 1.4　安裝成功

安裝好後，開啟 DOS 視窗，輸入 java –version 指令，如果出現如圖 1.5 所示的提示就代表 JDK 安裝成功了。

```
C:\Users\10138>java -version
java version "1.8.0_231"
Java(TM) SE Runtime Environment (build 1.8.0_231-b11)
Java HotSpot(TM) 64-Bit Server VM (build 25.231-b11, mixed mode)
```

圖 1.5　Java 版本編號

在 JetBrains 網 站 (https://www.jetbrains.com/idea/) 下 載 IntelliJ IDEA 2019. 2.4，如圖 1.6 所示。

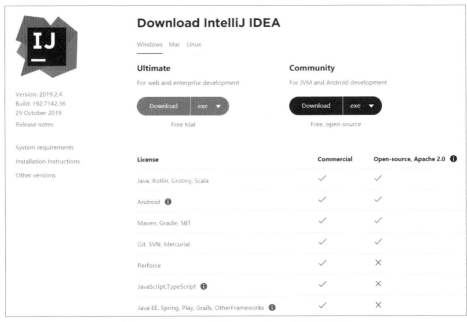

圖 1.6　IntelliJ IDEA 2019.2.4 的下載頁面

根據提示進行安裝，選擇安裝位置，如圖 1.7 所示。

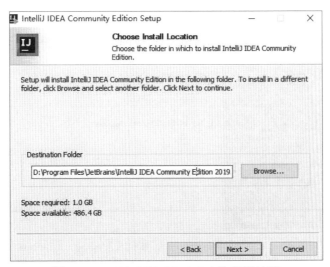

圖 1.7　IntelliJ IDEA 2019.2.4 的安裝位置

然後執行安裝過程,分析檔案,如圖 1.8 所示。

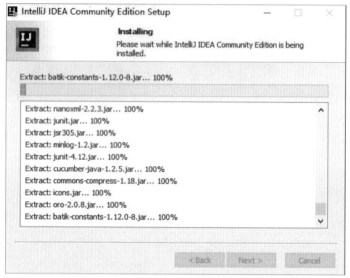

圖 1.8 IntelliJ IDEA 2019.2.4 的安裝過程

當出現如圖 1.9 所示的介面時表示安裝完成。

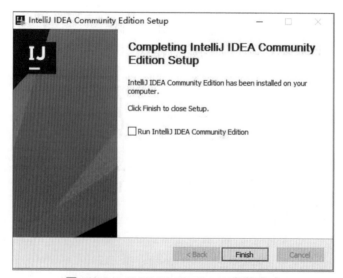

圖 1.9 IntelliJ IDEA 2019.2.4 安裝成功

1.3 在 Ubuntu 環境中架設 Kotlin 開發環境

在如圖 1.1 所示的介面和 IntelliJ IDEA 的網站下載對應的 Linux 平台的安裝檔案：jdk-8u231- linux-x64.tar.gz 和 ideaIC-2019.2.4.tar.gz。

執行以下指令將 jdk-8u231-linux-x64.tar.gz 解壓到 /opt 目錄，如圖 1.10 所示：

```
tar -xvf jdk-8u231-linux-x64.tar.gz
```

```
hadoop@ubuntu:/opt$ sudo tar -xvf jdk-8u231-linux-x64.tar.gz
jdk1.8.0_231/
jdk1.8.0_231/lib/
jdk1.8.0_231/lib/visualvm/
jdk1.8.0_231/lib/visualvm/visualvm/
jdk1.8.0_231/lib/visualvm/visualvm/.lastModified
jdk1.8.0_231/lib/visualvm/visualvm/modules/
jdk1.8.0_231/lib/visualvm/visualvm/modules/com-sun-tools-visualvm-uisupport.jar
jdk1.8.0_231/lib/visualvm/visualvm/modules/com-sun-tools-visualvm-application-views.jar
jdk1.8.0_231/lib/visualvm/visualvm/modules/com-sun-tools-visualvm-tools.jar
jdk1.8.0_231/lib/visualvm/visualvm/modules/com-sun-tools-visualvm-profiler.jar
jdk1.8.0_231/lib/visualvm/visualvm/modules/com-sun-tools-visualvm-jvmstat.jar
jdk1.8.0_231/lib/visualvm/visualvm/modules/com-sun-tools-visualvm-jmx.jar
jdk1.8.0_231/lib/visualvm/visualvm/modules/com-sun-tools-visualvm-core.jar
jdk1.8.0_231/lib/visualvm/visualvm/modules/com-sun-tools-visualvm-host-views.jar
jdk1.8.0_231/lib/visualvm/visualvm/modules/com-sun-tools-visualvm-heapdump.jar
jdk1.8.0_231/lib/visualvm/visualvm/modules/com-sun-tools-visualvm-attach.jar
jdk1.8.0_231/lib/visualvm/visualvm/modules/com-sun-tools-visualvm-application.jar
jdk1.8.0_231/lib/visualvm/visualvm/modules/com-sun-tools-visualvm-modules-appui.jar
jdk1.8.0_231/lib/visualvm/visualvm/modules/com-sun-tools-visualvm-charts.jar
jdk1.8.0_231/lib/visualvm/visualvm/modules/com-sun-tools-visualvm-coredump.jar
jdk1.8.0_231/lib/visualvm/visualvm/modules/locale/
jdk1.8.0_231/lib/visualvm/visualvm/modules/locale/com-sun-tools-visualvm-core_ja.jar
jdk1.8.0_231/lib/visualvm/visualvm/modules/locale/com-sun-tools-visualvm-host-views_ja.jar
jdk1.8.0_231/lib/visualvm/visualvm/modules/locale/com-sun-tools-visualvm-host_ja.jar
jdk1.8.0_231/lib/visualvm/visualvm/modules/locale/com-sun-tools-visualvm-application_ja.jar
jdk1.8.0_231/lib/visualvm/visualvm/modules/locale/com-sun-tools-visualvm-coredump_zh_CN.jar
jdk1.8.0_231/lib/visualvm/visualvm/modules/locale/com-sun-tools-visualvm-threaddump_zh_CN.jar
jdk1.8.0_231/lib/visualvm/visualvm/modules/locale/com-sun-tools-visualvm-host-remote_zh_CN.jar
jdk1.8.0_231/lib/visualvm/visualvm/modules/locale/com-sun-tools-visualvm-profiler_zh_CN.jar
jdk1.8.0_231/lib/visualvm/visualvm/modules/locale/com-sun-tools-visualvm-jmx_zh_CN.jar
jdk1.8.0_231/lib/visualvm/visualvm/modules/locale/com-sun-tools-visualvm-jvm_ja.jar
jdk1.8.0_231/lib/visualvm/visualvm/modules/locale/com-sun-tools-visualvm-profiling_zh_CN.jar
jdk1.8.0_231/lib/visualvm/visualvm/modules/locale/com-sun-tools-visualvm-host-views_zh_CN.jar
jdk1.8.0_231/lib/visualvm/visualvm/modules/locale/com-sun-tools-visualvm-attach_ja.jar
jdk1.8.0_231/lib/visualvm/visualvm/modules/locale/com-sun-tools-visualvm-jmx_ja.jar
jdk1.8.0_231/lib/visualvm/visualvm/modules/locale/com-sun-tools-visualvm-profiling_ja.jar
jdk1.8.0_231/lib/visualvm/visualvm/modules/locale/com-sun-tools-visualvm-profiler_ja.jar
jdk1.8.0_231/lib/visualvm/visualvm/modules/locale/com-sun-tools-visualvm-heapdump_zh_CN.jar
jdk1.8.0_231/lib/visualvm/visualvm/modules/locale/org-netbeans-core-windows_visualvm.jar
jdk1.8.0_231/lib/visualvm/visualvm/modules/locale/com-sun-tools-visualvm-jvm_zh_CN.jar
jdk1.8.0_231/lib/visualvm/visualvm/modules/locale/org-netbeans-modules-profiler_visualvm.jar
jdk1.8.0_231/lib/visualvm/visualvm/modules/locale/com-sun-tools-visualvm-charts_zh_CN.jar
jdk1.8.0_231/lib/visualvm/visualvm/modules/locale/com-sun-tools-visualvm-application_zh_CN.jar
jdk1.8.0_231/lib/visualvm/visualvm/modules/locale/com-sun-tools-visualvm-uisupport_ja.jar
jdk1.8.0_231/lib/visualvm/visualvm/modules/locale/com-sun-tools-visualvm-host-remote_ja.jar
jdk1.8.0_231/lib/visualvm/visualvm/modules/locale/com-sun-tools-visualvm-charts_ja.jar
jdk1.8.0_231/lib/visualvm/visualvm/modules/locale/org-netbeans-core_visualvm.jar
```

圖 1.10　jdk-8u231-linux-x64.tar.gz 的安裝過程

解壓完會產生一個目錄 jdk1.8.0_231/。然後在 .bashrc 檔案中設定環境變數，如下：

```
export JAVA_HOME=/opt/jdk1.8.0_231/
export JRE_HOME=${JAVA_HOME}/jre
export CLASSPATH=.:${JAVA_HOME}/lib:${JRE_HOME}/lib
export PATH=${JAVA_HOME}/bin:$PATH
```

執行 source .bashrc，使環境變數生效，然後執行 java -version，出現如圖 1.11 所示的內容，表示 Java 安裝成功。

```
hadoop@ubuntu:/opt$ java -version
java version "1.8.0_231"
Java(TM) SE Runtime Environment (build 1.8.0_231-b11)
Java HotSpot(TM) 64-Bit Server VM (build 25.231-b11, mixed mode)
```

圖 1.11 Java 版本編號

接下來安裝 IntelliJ IDEA，將 ideaIC-2019.2.4.tar.gz 解壓到 /opt 目錄，執行以下指令：

```
tar -xvf ideaIC-2019.2.4.tar.gz
```

安裝過程如圖 1.12 所示：

解壓後會產生一個新目錄 idea-IC-192.7142.36/。為了能夠在桌面上啟動 IntelliJ IDEA，需要建立捷徑。在 /usr/share/applications/ 目錄下建立 intellij-idea.desktop，將以下內容複製到 intellij-idea.desktop 中：

```
1.  [Desktop Entry]
2.  Name=IntelliJ IDEA
3.  Exec=/opt/idea-IC-192.7142.36/bin/idea.sh
4.  Comment=IntelliJ IDEA
5.  Icon=/opt/idea-IC-192.7142.36/bin/idea.png
6.  Type=Application
7.  Terminal=false
8.  Encoding=UTF-8
```

```
hadoop@ubuntu:/opt$ sudo tar -xvf ideaIC-2019.2.4.tar.gz
idea-IC-192.7142.36/LICENSE.txt
idea-IC-192.7142.36/NOTICE.txt
idea-IC-192.7142.36/bin/appletviewer.policy
idea-IC-192.7142.36/bin/idea.svg
idea-IC-192.7142.36/bin/log.xml
idea-IC-192.7142.36/build.txt
idea-IC-192.7142.36/lib/FastInfoset-1.2.15.jar
idea-IC-192.7142.36/lib/aapt-proto-jarjar.jar
idea-IC-192.7142.36/lib/annotations.jar
idea-IC-192.7142.36/lib/ant/CONTRIBUTORS
idea-IC-192.7142.36/lib/ant/INSTALL
idea-IC-192.7142.36/lib/ant/KEYS
idea-IC-192.7142.36/lib/ant/LICENSE
idea-IC-192.7142.36/lib/ant/NOTICE
idea-IC-192.7142.36/lib/ant/README
idea-IC-192.7142.36/lib/ant/WHATSNEW
idea-IC-192.7142.36/lib/ant/contributors.xml
idea-IC-192.7142.36/lib/ant/lib/README
idea-IC-192.7142.36/lib/ant/lib/ant-antlr.jar
idea-IC-192.7142.36/lib/ant/lib/ant-antlr.pom
idea-IC-192.7142.36/lib/ant/lib/ant-apache-bcel.jar
idea-IC-192.7142.36/lib/ant/lib/ant-apache-bcel.pom
idea-IC-192.7142.36/lib/ant/lib/ant-apache-bsf.jar
idea-IC-192.7142.36/lib/ant/lib/ant-apache-bsf.pom
idea-IC-192.7142.36/lib/ant/lib/ant-apache-log4j.jar
idea-IC-192.7142.36/lib/ant/lib/ant-apache-log4j.pom
idea-IC-192.7142.36/lib/ant/lib/ant-apache-oro.jar
idea-IC-192.7142.36/lib/ant/lib/ant-apache-oro.pom
idea-IC-192.7142.36/lib/ant/lib/ant-apache-regexp.jar
idea-IC-192.7142.36/lib/ant/lib/ant-apache-regexp.pom
idea-IC-192.7142.36/lib/ant/lib/ant-apache-resolver.jar
idea-IC-192.7142.36/lib/ant/lib/ant-apache-resolver.pom
idea-IC-192.7142.36/lib/ant/lib/ant-apache-xalan2.jar
idea-IC-192.7142.36/lib/ant/lib/ant-apache-xalan2.pom
idea-IC-192.7142.36/lib/ant/lib/ant-commons-logging.jar
idea-IC-192.7142.36/lib/ant/lib/ant-commons-logging.pom
idea-IC-192.7142.36/lib/ant/lib/ant-commons-net.jar
idea-IC-192.7142.36/lib/ant/lib/ant-commons-net.pom
idea-IC-192.7142.36/lib/ant/lib/ant-jai.jar
idea-IC-192.7142.36/lib/ant/lib/ant-jai.pom
idea-IC-192.7142.36/lib/ant/lib/ant-javamail.jar
idea-IC-192.7142.36/lib/ant/lib/ant-javamail.pom
idea-IC-192.7142.36/lib/ant/lib/ant-jdepend.jar
idea-IC-192.7142.36/lib/ant/lib/ant-jdepend.pom
idea-IC-192.7142.36/lib/ant/lib/ant-jmf.jar
idea-IC-192.7142.36/lib/ant/lib/ant-jmf.pom
idea-IC-192.7142.36/lib/ant/lib/ant-jsch.jar
```

圖 1.12　idealC-2019.2.4.tar.gz 的安裝過程

儲存後，執行 sudo chmod +x intellij-idea.desktop 指令。按 Windows 鍵，在
搜尋框中輸入 "int"，如圖 1.13 所示。

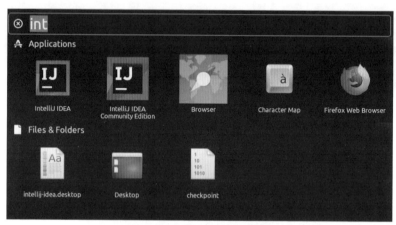

圖 1.13 IntelliJ IDEA 圖示

將 "IntelliJ IDEA" 這個圖示拖曳到桌面，這樣可以方便快速啟動 IntelliJ IDEA，如圖 1.14 所示。

雙擊圖示，啟動 IntelliJ IDEA，可以看到版本編號，表明 IntelliJ IDEA 安裝成功，如圖 1.15 所示。

圖 1.14 IntelliJ IDEA
的捷徑圖示

圖 1.15 IntelliJ IDE 安裝成功

1.4 在 macOS 環境中架設 Kotlin 開發環境

在如圖 1.1 所示的介面和 IntelliJ IDEA 的網站下載對應的 macOS 平台的安裝檔案:jdk-8u231-macosx-x64.dmg、ideaIC-2019.2.4.dmg。

雙擊 jdk-8u231-macosx-x64.dmg 安裝套件,開啟如圖 1.16 所示的視窗,按照系統提示進行安裝。

圖 1.16 jdk-8u231-macosx-x64.dmg 的安裝過程

然後就可以設定系統的 Java 環境變數了。開啟終端,進入目前使用者的 home 目錄,開啟 .bash_profile 並編輯:在檔案的尾端加入下面這行敘述並儲存。

```
export JAVA_HOME=/Library/Java/JavaVirtualMachines/jdk1.8.0_211.jdk/Contents/Home
```

雙擊 ideaIC-2019.2.4.dmg 安裝套件,待進度指示器讀完之後,拖曳 IDEA 圖示到 Applications,如圖 1.17 所示。

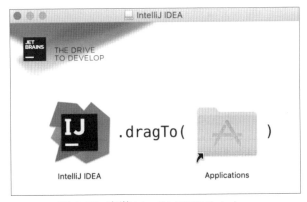

圖 1.17 安裝 ideaIC-2019.2.4.dmg

1.5 第一個 Kotlin 程式

本書使用 Windows 平台介紹。下面建立第一個 Kotlin 程式。首先開啟 IntelliJ IDEA，新增一個套件 io.kang.chapter01，然後新增一個 First KotlinApp.kt 檔案。Kotlin 程式的目錄如圖 1.18 所示。

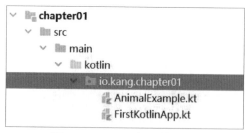

圖 1.18 Kotlin 程式的目錄

Kotlin 程式碼如下：

```
1.  package io.kang.chapter01        // 套件名稱
2.  fun main() {                     // 主函數
3.     println("Hello, World!")      // 列印 "Hello, World!"
4.  }
```

首先，定義一個套件。然後，定義程式的 main 函數，這是程式的入口。Kotlin 1.3 版本定義 main() 函數時可以不帶任何參數。main 函數沒有指定傳回類型，不傳回任何值。最後是方法區塊，println 方法輸出一個字串。

執行這個程式，可以輸出 "Hello，World!"，如圖 1.19 所示。

io.kang.chapter01.FirstKotlinAppKt ×
```
"D:\Program Files\Java\jdk1.8.0_171\bin\java.exe" ...
Hello, World!
```

圖 1.19 Kotlin 程式執行結果

1.6 小結

本章首先以一個尋找年齡最大的 Animal 物件的實例引用 Kotlin 語言,直觀地展示了 Kotlin 語言的特性——它是靜態類型語言、支援函數式程式設計和物件導向程式設計、具有很強的實用性、語法簡潔、安全性高、和 Java 語言互通性很好。

其次,本章圖文並茂地介紹了在三大主流平台上安裝 Java 1.8、IntelliJ IDEA 及 Kotlin 環境的過程,為後續章節撰寫 Kotlin 程式奠定了基礎。

最後,本章用 IntelliJ IDEA 撰寫了一個 Kotlin 程式,以讓大家對 Kotlin 有初步的了解。第 2 章將介紹函數式程式設計。

1.6 小結

函數式程式設計介紹

本章主要介紹函數式程式設計的概念、特點。函數式程式設計具有不變性、惰性求值、參考透明、無副作用等特點。本章將介紹 Java、Kotlin、Scala 語言的特點,並對它們的差異進行比較。

2.1 初識函數式程式設計

函數式程式設計以純函數的方式撰寫程式。純函數指的是一個函數在程式的執行過程中除了根據輸入參數列出運算結果外不受其他因素影響。純函數的核心目的是撰寫無副作用的程式,它有很多特性,包含不變性、惰性求值等。

函數式程式設計更加強調程式執行的結果而非執行的過程,宣導利用許多簡單的執行單元讓計算結果不斷演進,逐層推導複雜的運算,而非設計一個複雜的執行過程。下面透過兩個實例來比較一下指令式程式設計和函數式程式設計。

對一個陣列進行累加。指令式程式設計一個一個檢查元素，進行累加，程式如下：

```java
1.  public class AddDemo {
2.      // 宣告陣列
3.      public final static List<Integer> nums = Arrays.asList(0, 1, 2, 3, 4,
    5, 6, 7, 8, 10);
4.      // 加和方法
5.      public static Integer sum(List<Integer> nums) {
6.          int result = 0;
7.          for (Integer num : nums) {
8.              result += num;
9.          }
10.         return result;
11.     }
12.     // 主函數
13.     public static void main(String[] args) {
14.         System.out.println(AddDemo.sum(nums));
15.     }
16. }
```

使用函數式程式設計，設定初值和運算方法（累加），檢查集合，即可獲得累加和。函數式程式設計更加關注運算的結果：

```kotlin
1.  fun main() {
2.      // 宣告陣列
3.      val nums = listOf(0, 1, 2, 3, 4, 5, 6, 7, 8, 10)
4.      // 列印陣列元素的和
5.      println(nums.fold(0, { acc, i ->  acc + i}))
6.  }
```

求費氏數列，指令式程式設計的程式如下，按照費氏數列的定義一步步計算結果：

```java
1.  public class FibonacciDemo {
2.      // 費氏數列方法
3.      public static int fibonacci(int number) {
```

```
4.          if (number == 1) {
5.              return 1;
6.          }
7.          if(number == 2) {
8.              return 2;
9.          }
10.         int a = 1;
11.         int b = 2;
12.         for(int cnt = 3; cnt <= number; cnt++) {
13.             int c = a + b;
14.             a = b;
15.             b = c;
16.         }
17.         return b;
18.     }
19.     // 主函數
20.     public static void main(String[] args) {
21.         System.out.println(FibonacciDemo.fibonacci(5));
22.     }
23. }
```

求費氏數列，函數式程式設計的程式如下所示，使用函數式程式設計將每一步拆解為一個函數，透過函數的組合獲得最後結果：

```
1.  // 費氏數列函數
2.  fun fibonacci(num: Int): Int {
3.      return Stream.iterate(listOf(1, 1), {
4.              t ->  listOf(t[1], t[0] + t[1])
5.          })
6.          .limit(num.toLong())
7.          .map { n -> n[1] }
8.          .toList()[num - 1]
9.  }
10. // 主函數
11. fun main() {
12.     print(fibonacci(8))
13. }
```

2.2 函數式程式設計的特點

函數式程式設計具有以下特點。

函數是一等公民：函數與其他資料類型一樣，處於平等地位，不但可以將函數作為參數、傳回值使用，而且可以用函數名稱進行參考，甚至可以匿名呼叫。下面的實例展示了函數作為參數、傳回值、匿名函數的用法。

```
1.  fun main() {
2.      val arr = listOf(1, 2, 3, 4)
3.      // 作為參數
4.      println(arr.reduce { acc, i -> acc + i })
5.      // 作為傳回值
6.      fun operate(num: Int): (Int) -> Int {
7.          return {num -> num * 2}
8.      }
9.      // 匿名函數
10.     fun(x: Int, y: Int) = x + y
11. }
```

參考透明，無副作用：函數的執行不依賴外部變數或「狀態」，只依賴輸入的參數，任何時候只要參數相同，參考函數所得到的傳回值總是相同的。副作用是指有超出函數控制的操作，例如在執行過程中操作檔案系統、資料庫等外部資源。沒有副作用使得函數式程式設計各個獨立部分的執行順序可以被隨意打亂，多個執行緒之間不共用狀態，不會造成資源爭用，不需要用鎖來保護可變狀態，不會出現鎖死，這樣可以更進一步地進行無鎖的平行處理操作。

在下面的實例中，xs 是不可變物件，多次呼叫 xs.slice 方法，導入參數相同，傳回結果相同。xs1 是可變物件，多次呼叫 xs1.add 方法，會改變 xs1 的值，產生副作用。

```
1.  fun main() {
```

```
2.      val xs = listOf(1, 2, 3, 4, 5)
3.      // 多次呼叫 xs.slice 方法的傳回結果一致
4.      println(xs.slice(IntRange(0, 3)))
5.      println(xs.slice(IntRange(0, 3)))
6.      println(xs.slice(IntRange(0, 3)))
7.
8.      var xs1 = mutableListOf(1, 2, 3)
9.      // 多次呼叫 xs1.add 方法的傳回結果不一樣
10.     xs1.add(4)
11.     println(xs1)
12.     xs1.add(5)
13.     println(xs1)
14.     xs1.add(6)
15.     println(xs1)
16. }
```

高階函數：可以將函數作為參數進行傳遞，也可以在函數內部宣告一個函數，那麼外層的函數就被稱作高階函數。

```
1.  // 函數 add 作為參數傳遞
2.  fun test(a: Int , b: Int, add: (num1 : Int , num2 : Int) -> Int) : Int{
3.      return add(a, b)
4.  }
5.  fun main() {
6.      // 呼叫 test 函數
7.      println(test(10, 11) { num1: Int, num2: Int -> num1 + num2})
8.  }
```

柯里化（Currying）：把接收多個參數的函數轉換成接收單一參數的函數，傳回新函數，該函數接收剩餘的參數並且傳回結果。柯里化的好處是減少了函數的參數個數，並且模組化了每一步計算，與設計模式中的轉接器模式（將一個介面轉為另一個介面）類似。並且柯里化的應用之一——惰性求值，也是函數式程式設計的重要特性。惰性求值在將運算式設定值給變數（或稱作綁定）時並不計算運算式的值，而是在變數第一次被使用時才進行計算。

下面的實例展示了柯里化、惰性求值。add(1) 輸入一個參數，傳回一個函數；add(1)(2)，輸入兩個參數，傳回一個函數；add(1)(2)(3) 輸入三個參數，傳回一個結果。惰性求值在呼叫時才進行初始化，println(y) 方法不會初始化 y，println(y.values) 方法才會初始化 y。

```
1.  fun main() {
2.      // 定義一個柯里化函數 add,
3.      // 接收參數 a,傳回一個函數
4.      // 接收參數 b,傳回一個函數
5.      // 接收參數 c,傳回一個結果
6.      fun add(a: Int): (b: Int) -> (c: Int) -> Int {
7.          return {b -> {c ->  a + b * c} }
8.      }
9.      // 呼叫函數 add
10.     println(add(1)(2)(3))
11.     // 惰性載入
12.     val y = lazy {
13.         println("y")
14.         13
15.     }
16.     // 此時,13 沒有產生實體
17.     println(y)
18.     // 此時,13 產生實體
19.     println(y.value)
20. }
```

2.3 Scala、Kotlin、Java 的比較

Scala、Kotlin、Java 都是以 Java 虛擬機器（JVM）為目標執行環境的語言。Scala、Kotlin 執行於 JVM 上，所以 Scala、Kotlin 可以存取 Java 類別庫，能夠與 Java 架構進行互動操作。Scala、Kotlin 將物件導向和函數式程式設計有機結合在一起，既有動態語言的靈活性和簡潔性，也有靜態語言的類型檢查功能來保障安全和加強執行效率。

Scala 是一種多範式的程式語言。Scala 具有物件導向的特性，每個值都是物件，物件的資料類型及行為由類別和特性描述。類別抽象機器制的擴充有兩種途徑：一種途徑是子類別繼承，另一種途徑是靈活的混入機制。這兩種途徑能避免多重繼承的種種問題。Scala 是一種函數式語言，其函數也能當成值來使用。Scala 提供了輕量級的語法，用於定義匿名函數，支援高階函數，允許巢狀結構多層函數，並支援柯里化。可以利用 Scala 的模式比對撰寫類似正規表示法的程式處理 XML 資料。Scala 具備類型系統，編譯成功時檢查，可保障程式的安全性和一致性。類型系統支援泛型類別、協變和逆變、標記、類型參數的上下限約束、把類別和抽象類型作為物件成員、複合類型、參考自己時顯性地指定類型、視圖、多形方法、擴充性等。Scala 使用 Actor 作為其平行處理模型，Actor 是類似執行緒的實體，透過電子郵件收發訊息。Actor 可以重複使用執行緒，因此在程式中可以使用數百萬個 Actor，而執行緒只能建立數千個。

Kotlin 語法和 Java 很像，容易上手，推薦以循序漸進的方式開發專案：允許專案中同時存在 Java 和 Kotlin 程式檔案，允許 Java 與 Kotlin 互相呼叫。這使得開發者可以很方便地在已有專案中引用 Kotlin。Kotlin 程式量少且程式尾端沒有分號。Kotlin 是空安全的，在編譯時期就處理了各種 null 的情況，避免了即時執行異常。Kotlin 可擴充函數，可以擴充任意類別的更多的特性。Kotlin 也是函數式的，可使用 Lambda 運算式來更方便地解決問題。Kotlin 具有高度互通性，可以繼續使用所有用 Java 撰寫的程式和函數庫，甚至可以在一個專案中使用 Kotlin 和 Java 兩種語言混合程式設計。

Java 是企業級的開發語言，很多網際網路公司會使用 Java 開發後台應用，目前主要使用 Java 7 和 Java 8 進行開發。Java 是物件導向的、分散式的、穩固的、安全的，近些年也增加了一些函數式語言的特性。Java 8 中引用了 Lambda 運算式、Stream，提供了類似 Scala、Kotlin 的函數式運算。Java 8 使用 Optional 避免空指標；Java 11 中引用了本機變數類型推斷，增強了 Stream，增強了 Optional。

Kotlin 是更好的 Java，Kotlin 是一種實用的語言，旨在加強 Java 開發人員的工作效率，在 Java 模型之上有一些 Scala 特性。它增加了 Java 程式設計師想要的功能，如 Lambda 運算式和基本功能。雖然 Kotlin 為函數式程式設計提供了一些支援，但它確實可以實現更簡單的程序式程式設計或指令式程式設計。

Scala 比 Java 更強大，旨在完成 Java 無法做到的事情。Scala 為進階函數程式設計提供了很好的支援，這增加了複雜性，使 Scala 成為一種比較難學習的語言。Scala 中有多種程式設計風格，可能會導致混亂或為每種需求提供最佳風格，這會導致更高的開發成本。

Kotlin 可以很方便地融入 Java 已有的生態，如 Spring Boot、Spring Cloud、Dubbo 等。Spring 5 引用了對 Kotlin 的支援。Scala 有自己的生態，目前其比較流行的 Web 架構是 Lift 架構和 Play 架構，微服務架構有 Akka，響應式微服務架構為 Lagom Framework。Scala 的架構有一定的學習門檻和難度，對普通 Java 開發者來講不太容易掌握。

基於以上比較，Kotlin 和 Java 良好的互通性使得 Java 開發者可以很快地掌握使用 Kotlin 開發微服務應用的方法，這也是本書的目標。

2.4 小結

本章介紹了函數式程式設計及其特點，函數是一等公民，函數式程式設計參考透明、無副作用，支援高階函數，可柯里化並進行惰性求值等。本章還比較了 Scala、Kotlin、Java 三種程式語言的差異，Scala 具有自動轉型、隱式參數等很多 Kotlin 沒有的特性；Spring 5 對 Kotlin 支援較好，可以利用 Java 的生態元件；Java 新增了 Lambda 運算式、類型推斷、不可變集合等特性，逐漸向函數式程式設計接近。

Chapter 03

Kotlin 的語法

本章介紹 Kotlin 的語法，包含基礎語法：基底資料型態、套件名稱和
參考、流程控制、傳回和跳躍；類別：介面、特殊類別、泛型、委
派；函數和 Lambda 運算式；集合：集合類型、集合操作；程式碼協同：
程式碼協同基礎知識及進階。透過本章的介紹，讀者對 Kotlin 的語法會有
直觀且快速的了解。

3.1 基礎語法

本節主要介紹 Kotlin 使用的基底資料型態、套件名稱和參考、流程控制、
傳回和跳躍等基礎語法。

3.1.1 基底資料型態

Kotlin 使用的基底資料型態包含：數值型、字元型、布林型、陣列和字串。

數值型分為整數和浮點數。整數有 8 種類型，如表 3.1 所示。

表 3.1 Kotlin 中整數的類型

類型	位元數 (bit)	最小值	最大值
Byte	8	-128	127
Short	16	-32,768	32,767
Int	32	-2,147,483,648 (-2^{31})	2,147,483,647 (2^{31}-1)
Long	64	-9,223,372,036,854,775,808 (-2^{63})	9,223,372,036,854,775,807 (2^{63}-1)
UByte	8	0	255
UShort	16	0	65,535
UInt	32	0	$2^{32} - 1$
ULong	64	0	$2^{64} - 1$

浮點數有 Float 和 Double 兩種類型，如表 3.2 所示。

表 3.2 Kotlin 中浮點數的類型

類型	位元數 (bit)	有效位數	指數位數	小數位數
Float	32	24	8	6 或 7
Double	64	53	11	15 或 16

Kotlin 不允許自動轉型類型。舉例來說，方法 printDouble() 的參數是 Double 類型的，呼叫該方法時，參數必須是 Double 類型的。每種類型都提供了以下顯性轉換方法。

- toByte(): Byte
- toShort(): Short
- toInt(): Int
- toLong(): Long
- toFloat(): Float
- toDouble(): Double
- toChar(): Char

```
1.  fun main() {
2.      // 列印數值
3.      fun printDouble(d: Double) { print(d) }
4.      // 宣告變數
5.      val i = 1
6.      val d = 1.1
7.      val f = 1.1f
8.
9.      printDouble(d)
10.     printDouble(i)   // 錯誤：類型不符合
11.     printDouble(f)   // 錯誤：類型不符合
12. }
```

注意：Long 類型的數字的結尾有 L，如 123L。十六進位數以 0X 開頭，如 0X0F。二進位數字以 0b 開頭，如 0b00001011。Float 類型的數字的結尾有 f 或 F。無號數值，以 u 或 U 結尾。無號長整數以 uL 或 UL 結尾。

數值也支援用 "_" 增強可讀性，以下面的程式所示：

```
1.  // 數值使用 "_" 增強可讀性
2.  val oneMillion = 1_000_000
3.  val creditCardNumber = 1234_5678_9012_3456L
4.  val socialSecurityNumber = 999_99_9999L
5.  val hexBytes = 0xFF_EC_DE_5E
6.  val bytes = 0b11010010_01101001_10010100_10010010
```

注意：這些數值型態可為空時（舉例來說，Int?），數值是包裝類別，其他情況下的數值是基底資料型態。包裝類別的值相等，但是記憶體位址不同，以下面的程式所示：

```
1.  // a 是包裝類別，== 比較的是值是否相等，=== 比較的是位址是否相等
2.  val a: Int = 10000
3.  println(a === a) // 列印 'true'
4.  val boxedA: Int? = a
5.  val anotherBoxedA: Int? = a
6.  println(boxedA === anotherBoxedA)    // !!! 列印 'false'!!!
7.  println(boxedA == anotherBoxedA)     // 列印 'true'
```

數值型態支援使用 ==、!=、>、<、>=、<= 進行範圍判斷。此外，Int 和 Long 類型的數值支援以下位元運算。

- shl(bits)──有號數值左移
- shr(bits)──有號數值右移
- ushr(bits)──無號數值右移
- and(bits)──逐位元與
- or(bits)──逐位元或
- xor(bits)──逐位元互斥
- inv()──逐位元反轉

```
// in 的用法，表示範圍
val x = 1.01
val isInRange = x in 1.0..2.0
val isNotInRange = x !in 0.0..1.0
println(isInRange)      // 列印 'true'
println(isNotInRange)   // 列印 'true'
```

字元類型的關鍵字是 Char。字元類型用單引號，如 'a'。字元類型不能直接轉為數值類型，需要進行顯性轉換。此外，如果字元類型可為空，其是包裝類別，而非基底資料型態。

布林類型的關鍵字是 Boolean，只有兩個設定值 true 和 false。如果布林類型可為空，其也是包裝類別。布林類型支援 &&、||、! 三種運算子。

陣列類型的關鍵字是 Array，可以用 arrayOf() 建立一個陣列，舉例來説，arrayOf(1,2,3) 表示陣列 [1,2,3]。陣列有 get() 和 set() 方法，可以用 [] 替代這兩個方法，舉例如下：

```
1.  // 陣列類型舉例
2.  val arrayTemp = arrayOf(1,2,3)
3.  arrayTemp[0] = 0
4.  println(arrayTemp[0])      // 列印 '0'
5.  println(arrayTemp.size)    // 列印 '3'
```

Kotlin 提供 byteArray、shortArray 和 intArray 等用於建立不同類型的陣列。

```
1.  // 基底資料型態陣列舉例
2.  val intArray: IntArray = intArrayOf(1, 2, 3)
3.  val byteArray: ByteArray = byteArrayOf(1, 2, 3)
4.  val shortArray: ShortArray = shortArrayOf(1, 2, 3)
5.  val longArray: LongArray = longArrayOf(1, 2, 3)
6.  val charArray: CharArray = charArrayOf('1', '2', '3')
7.  val floatArray: FloatArray = floatArrayOf(1.0f, 2.0f, 3.0f)
8.  val doubleArray: DoubleArray = doubleArrayOf(1.0, 2.0, 3.0)
9.  val booleanArray: BooleanArray = booleanArrayOf(true, false, true)
```

String 表示字串。字串是不可變物件，可以用 [i] 檢查每一個字元。

```
1.  // 檢查字串
2.  val str = "hello"
3.  for (c in str) {
4.      println(c)
5.  }
```

可以用 "+" 連接字串：

```
1.  // 連接字串
2.  val s = "abc" + 1
3.  println(s + "def")
```

字串可以由 """ 分割，包含分行符號：

```
1.  // 使用 """ 初始化字串，包含分行符號
2.  val text = """
3.      for (c in "foo")
4.      print(c)
5.      """
```

在上面這個實例中，每行都輸出了很多空格，可以用 trimMargin() 方法去除空格，每行都有一個 "|" 字首，也可以自訂字首，如 ">"，呼叫 trimMargin(">") 即可：

```
1.  // 使用 trimMargin() 方法去除每行字串前的空格
2.  val text1 = """
3.  |Tell me and I forget.
4.  |Teach me and I remember.
5.  |Involve me and I learn.
6.  |(Benjamin Franklin)
7.  """.trimMargin()
```

Kotlin 的 String 支援範本，在字串中使用 "$" 可嵌入範本：

```
1.  // $ 支援字串範本
2.  val strTemplate = "abc"
3.  // 列印結果："abc.length is 3"
4.  println("$strTemplate.length is ${strTemplate.length}")
```

如果需要單獨使用 $，可以參考以下程式來實現：

```
1.  // 字串範本單獨使用 $
2.  val price = """
3.      |${'$'}9.99
4.      """.trimMargin()
5.  println(price) //prints $9.99
```

3.1.2 套件名稱和參考

Kotlin 的原始程式中一般都有一個套件名稱，所有的類別和方法等都在這個套件下：

```
1.  package io.kang.chapter03.basicsyntax
2.  // 方法
3.  fun printMessage() { /*...*/ }
4.  // 類別
5.  class Message { /*...*/ }
```

在上面這個實例中，printMessage 的全名是 io.kang.chapter03.basicsyntax. printMessage，Message 的全名是 io.kang.chapter03.basicsyntax.Message。

每一個 Kotlin 檔案在 JVM 平台中會預設引用以下套件：

- kotlin.*
- kotlin.annotation.*
- kotlin.collections.*
- kotlin.comparisons.*（從 1.1 版本開始）
- kotlin.io.*
- kotlin.ranges.*
- kotlin.sequences.*
- kotlin.text.*
- java.lang.*
- kotlin.jvm.*

我們可以引用一個類別，也可以引用套件中的所有內容，為了避免命名衝突，還可以給參考起別名：

```
1.  // 參考套件
2.  import io.kang.chapter03.basicsyntax.Message
3.  import io.kang.chapter03.basicsyntax.*
4.  import io.kang.chapter03.basicsyntax.Message as Message1
```

3.1.3 流程控制

Kotlin 的流程控制敘述有 if、when、for、while。if 的分支可以是程式區塊，最後的運算式作為該區塊的值。

```
1.  // 傳統用法
2.  var maxVal = 1
3.  if (1 < 2) maxVal = 2
4.
5.  // 使用 else
6.  if (1 > 2) {
```

```
7.      maxVal = 1
8.  } else {
9.      maxVal = 2
10. }
11. // maxVal 作為運算式
12. maxVal = if (1 > 2) 1 else 2
13. maxVal = if (1 > 2) {
14.     print("Choose 1")
15.     1
16. } else {
17.     print("Choose 2")
18.     2
19. }
```

when 敘述類似 switch，when 將它的參數與所有的分支條件順序比較，直到某個分支滿足條件。when 既可以被當作運算式使用，也可以被當作敘述使用。

```
1.  // when 用法舉例
2.  when (0) {
3.      is Int -> println("x is Int")
4.      else -> println("other type")
5.  }
```

for 敘述可以對任何提供反覆運算器（iterator）的物件進行檢查。

```
1.  // for 昇冪檢查範圍
2.  for (i in 1..3) {
3.      println(i)
4.  }
5.  // for 降冪檢查範圍
6.  for (i in 6 downTo 0 step 2) {
7.      println(i)
8.  }
9.  // for 檢查陣列
10. val array = arrayOf(1,2,3)
```

```
11. for (i in array.indices) {
12.     println(array[i])
13. }
14. // for 解析陣列的索引和元素
15. for ((index, value) in array.withIndex()) {
16.     println("the element at $index is $value")
17. }
```

While、do…while 在 Kotlin 中的用法和在 Java 中相同。

3.1.4 傳回和跳躍

Kotlin 有以下三種跳躍運算式：

- return——預設從最直接包圍它的函數或匿名函數傳回。
- break——終止最直接包圍它的循環。
- continue——繼續下一次最直接包圍它的循環。

在 Kotlin 中，任何運算式都可以用標籤（label）來標記。標籤的格式為在識別符號後跟 @ 符號，例如 abc@。可以用標籤限制 return、break、continue，舉例如下：

```
1.  // return 舉例
2.  fun foo() {
3.      listOf(1, 2, 3, 4, 5).forEach {
4.          if (it == 3) return   // 直接傳回到 foo() 的呼叫處
5.          print(it)
6.      }
7.      println("this point is unreachable")
8.  }
9.  // 使用 break 跳躍到 loop@
10. fun foo1() {
11.     loop@ for (i in 1..5) {
12.         for (j in 1..5) {
```

```
13.            if (j == 2) break@loop
14.            println("$i---$j")
15.        }
16.    }
17. }
18. // 使用 continue 跳躍到 loop@
19. fun foo2() {
20.    loop@ for (i in 1..5) {
21.        for (j in 1..5) {
22.            if (j == 2) continue@loop
23.            println("$i---$j")
24.        }
25.    }
26. }
```

3.2 類別

本節介紹如何定義 Kotlin 中的類別、屬性、介面，以及特殊類別、泛型、委派等語法的概念和用法。

3.2.1 類別、屬性、介面

在 Kotlin 中，使用關鍵字 class 宣告類別。類別宣告由類別名稱、類別頭（指定其類型參數、主建構函數等）及由大括號包圍的類別本體組成，其中類別本體中包含建構函數、初始化區塊、函數、屬性、巢狀結構類別與內部類別、物件宣告。類別頭與類別本體都是可選的；如果一個類別沒有類別本體，可以省略大括號。Kotlin 中的類別可以有一個主建構函數及一個或多個次建構函數。如果主建構函數沒有任何註釋或可見性修飾符號，可以省略 constructor 關鍵字。主建構函數不能包含任何程式。初始化程式可以放到以 init 關鍵字作為字首的初始化區塊（initializer block）中。

主建構函數中宣告的屬性可以是可變的（var）或唯讀的（val），建構函數的屬性可以有預設值。類別也可以宣告字首為 constructor 的次建構函數。如果類別有一個主建構函數，每個次建構函數需要委派給主建構函數，可以直接委派或透過其他次建構函數進行間接委派。委派到同一個類別的另一個建構函數用 this 關鍵字即可。所有初始化區塊中的程式都會在次建構函數本體之前執行。即使該類別沒有主建構函數，這種委派仍會隱式發生，並且仍會執行初始化區塊。

```kotlin
1.  package io.kang.chapter03.basicsyntax
2.  // 定義 Person 類別
3.  class Person(val name: String, var age: Int = 0) {
4.      var children: MutableList<Person> = mutableListOf<Person>();
5.      // 初始化區塊
6.      init {
7.          println("name is $name, age is $age")
8.      }
9.      // 次建構函數
10.     constructor(name: String, age: Int, parent: Person) : this(name, age) {
11.         parent.children.add(this)
12.     }
13. }
14. fun main() {
15.     // 產生實體一個 Person 物件，使用主建構函數
16.     val person = Person("yuan")
17.     // 產生實體一個 Person 物件，使用次建構函數
18.     val parentPerson = Person("kang", 1, person)
19. }
```

要建立一個類別的實例，可以像普通函數一樣呼叫建構函數。Kotlin 中沒有 new 關鍵字，Kotlin 中的所有類別都有一個共同的超類別 Any，其對沒有超型態宣告的類別來說是預設超類別。Any 有三個方法：equals()、hashCode() 與 toString()，因此，為所有 Kotlin 類別都定義了這些方法。

如果類別要被繼承，需要用 open 關鍵字修飾。Kotlin 對可覆蓋的成員（我們稱之為開放）及覆蓋後的成員需要進行顯性修飾。如果衍生類別有一個主建構函數，那麼其基礎類別類型可以（並且必須）用基礎類別的主建構函數參數就地初始化。如果衍生類別沒有主建構函數，那麼每個次建構函數必須使用 super 關鍵字初始化其基礎類別類型，或委派給另一個建構函數做到這一點。

```
1.  // 定義基礎類別 Base，open 為關鍵字，表示 Base 可以被繼承
2.  open class Base(val p: Int) {
3.      // open 修飾的屬性，方法可以被子類別覆蓋
4.      open val vertexCount: Int = 0
5.      open fun draw() {
6.          println("Base.draw()")
7.      }
8.      constructor(base: String, p: Int): this(p){
9.          println("base is $base")
10.     }
11. }
12. // 定義 Derived，它繼承了 Base
13. class Derived: Base {
14.     override var vertexCount = 4
15.     constructor(base: String, p: Int): super(base, p)
16.
17.     override fun draw() {
18.         println("Derived.draw()")
19.     }
20. }
21. fun main() {
22.     // 產生實體 Derived 類別
23.     val derived = Derived("kang", 1)
24.     derived.draw()
25. }
```

Derived.draw() 函數必須加上 override 修飾符號，如果沒加，編譯器將顯示出錯。如果函數沒有標記 open，那麼子類別中不允許定義相同簽名的函

數。將 open 修飾符號增加到 final 類別（即沒有 open 的類別）的成員上會不有作用。標記為 override 的成員本身是開放的，也就是說，它可以在子類別中被覆蓋。如果你想禁止再次被覆蓋，可使用 final 關鍵字。

類別及其中的某些成員可以被宣告為 abstract。抽象成員在本類別中可以不用實現，可以用一個抽象成員覆蓋一個非抽象的開放成員。

Kotlin 類別中的屬性既可以用關鍵字 var 宣告為可變的，也可以用關鍵字 val 宣告為唯讀的。要使用一個屬性，只要用名稱參考它即可。宣告一個屬性的完整語法如下：

```
var <propertyName>[: <PropertyType>] [= <property_initializer>]
    [<getter>]
    [<setter>]
1.  var allByDefault: Int? // 錯誤：需要進行顯性初始化，提供預設的 getter
                           和 setter 方法
2.  var initialized = 1    // 類型為 Int，提供預設的 getter 和 setter 方法
3.  val simple: Int?  // 類型為 Int，提供預設的 getter 方法必須在建構函數中初始化
4.  val inferredType = 1   // 類型為 Int，提供預設的 getter 方法
```

可以自訂 setter、getter 方法，舉例如下：

```
1.  class Demo(val aList: MutableList<String>) {
2.      // 定義 size 的 getter 和 setter
3.      var size: Int = 0
4.          get() = aList.size
5.          set(value) {
6.              field = value
7.          }
8.      // 定義 listToString 的 getter 和 setter
9.      var listToString: String
10.         get() = aList.toString()
11.         set(value) {
12.             aList.add(value)
13.         }
```

```
14. }
15. // 產生實體 Demo 物件，呼叫屬性的 getter 方法
16. val demo = Demo(MutableList<String>(10){"a"})
17. println(demo.size)
18. println(demo.listToString)
```

常數可以用 const 修飾。如果要推後初始化屬性和變數，可以用 lateinit 關鍵字。

Kotlin 的介面可以既包含抽象方法的宣告也包含實作方式。與抽象類別不同的是，介面無法儲存狀態。它可以有屬性但必須宣告為抽象或提供存取器實現，可使用關鍵字 interface 來定義介面。

```
1.  // 定義介面 Named
2.  interface Named {
3.      val name: String
4.  }
5.  // 定義介面 People，繼承 Named
6.  interface People : Named {
7.      val firstName: String
8.      val lastName: String
9.      // name 屬性的預設實現
10.     override val name: String get() = "$firstName $lastName"
11. }
12. // Employee 實現 People 介面
13. class Employee(
14.         // 不必實現 "name"，介面有預設實現
15.         override val firstName: String,
16.         override val lastName: String
17. ) : People
```

類別、物件、介面、建構函數、方法、屬性和它們的 setter 都可以有可見性修飾符號。Kotlin 有四個可見性修飾符號：private、protected、internal 和 public，預設可見性是 public。private 只在類別內部（包含其所有成

員）可見；protected 在子類別中可見。能見到類別宣告的本模組內的任
何用戶端都可見其 internal 成員；能見到類別宣告的任何用戶端都可見其
public 成員。如果你覆蓋一個 protected 成員並且沒有顯性指定其可見性，
那麼該成員還會是 protected 的。

```
1.  // 檔案名稱：classes.kt
2.  package io.kang.chapter03.basicsyntax
3.
4.  private fun foo() { ... } // 在 classes.kt 內可見
5.
6.  public var bar: Int = 5     // 該屬性隨處可見
7.      private set             // setter 只在 classes.kt 內可見
8.
9.  internal val baz = 6        // 相同模組內可見
```

3.2.2 特殊類別

Kotlin 中的特殊類別有以下幾種。

資料類別：我們經常建立一些只儲存資料的類別，用 data 標記。資料類別
的主建構函數需要至少有一個參數；所有參數標記為 val 或 var；資料類別
不能是抽象、開放、密封或內部的。如果產生的類別需要含有一個無參的
建構函數，則所有的屬性必須指定預設值。

```
1.  // 定義資料類別 User
2.  data class User(val name: String = "", val age: Int = 0)
3.  // 產生實體 User
4.  val jack = User(name = "Jack", age = 1)
5.  val olderJack = jack.copy(age = 2)
6.  // 解構 jack 的 name、age 屬性
7.  val (name, age) = jack
8.  println("$name, $age years of age")     // 輸出 "Jane, 35 years of age"
```

密封類別：用來表示受限的類別繼承結構，即一個值為有限的幾種類型且不能有任何其他類型。密封類別的子類別是包含狀態的多個實例。密封類別需要在類別名稱前面增加 sealed 修飾符號。所有子類別都必須在與密封類別本身相同的檔案中宣告。密封類別是抽象的，其建構函數預設是 private 的。

```
1.    // 密封類別 Expr
2.    sealed class Expr
3.    // Const、Sum 繼承 Expr
4.    data class Const(val number: Double) : Expr()
5.    data class Sum(val e1: Expr, val e2: Expr) : Expr()
6.    // 使用 when 進行類型判斷
7.    fun eval(expr: Expr): Double = when(expr) {
8.        is Const -> expr.number
9.        is Sum -> eval(expr.e1) + eval(expr.e2)
10.       // 不再需要 else 子句，因為我們已經覆蓋了所有的情況
11. }
```

巢狀結構類別：巢狀結構在其他類別內部的類別。標記為 inner 的巢狀結構類別能夠存取其外部類別的成員。

列舉類別：其最基本的用法是實現類型安全的列舉，每一個列舉都可以初始化。列舉常數還可以宣告其帶有對應的方法及覆蓋了基礎類別方法的匿名類別。

```
1.    // 列舉類別
2.    enum class ProtocolState {
3.        WAITING {
4.            override fun signal() = TALKING
5.        },
6.        TALKING {
7.            override fun signal() = WAITING
8.        };
9.
```

```
10.     abstract fun signal(): ProtocolState
11. }
```

3.2.3 泛型

Kotlin 中的類別也可以有類型參數。Kotlin 的類型系統有宣告處型變
（declaration-site variance）與類型投影（type projection）。

```
1.  // 定義一個泛型類別 Box
2.  class Box<T>(t: T) {
3.      var value = t
4.  }
5.  // 產生實體時，設定類型為 Int
6.  val box: Box<Int> = Box<Int>(1)
```

Kotlin 用 out 關鍵字表示泛型是協變類型的。當一個類別 C 的類型參數 T
被宣告為 out 時，它就只能出現在 C 的成員的輸出位置，這樣，C<Base>
可以安全地作為 C<Derived> 的超類別。類別 C 在參數 T 上是協變的，或
說 T 是一個協變的類型參數。C 是 T 的生產者，而非 T 的消費者。out 修
飾符號也被稱為型變註釋，並且由於它在類型參數宣告處提供，所以我們
稱之為宣告處型變。

```
1.  // 定義協變類型
2.  interface Source<out T> {
3.      fun nextT(): T
4.  }
5.
6.  fun demo(strs: Source<String>) {
7.      val objects: Source<Any> = strs    // 這個沒問題，因為 T 是一個 out 參數
8.  }
```

與 out 相反，Kotlin 又補充了一個型變註釋：in。它使得一個類型參數逆
變：只可以被消費而不可以被生產。

```
1.  // 定義逆變類型
2.  interface Comparable<in T> {
3.      operator fun compareTo(other: T): Int
4.  }
5.  fun demo(x: Comparable<Number>) {
6.      x.compareTo(1.0) // 1.0 擁有類型 Double，它是 Number 的子類型
7.      // 因此，我們可以將 x 指定給類型為 Comparable <Double> 的變數
8.      val y: Comparable<Double> = x   // 正確！
9.  }
```

將類型參數 T 宣告為 out 非常方便，但是有些類別實際上不能被限制為只傳回 T。例如 Array，該類別在 T 上既不能是協變的也不能是逆變的。

```
1.  // Array 在 T 上既不能是協變的，也不能是逆變的
2.  class Array<T>(val size: Int) {
3.      fun get(index: Int): T {  }
4.      fun set(index: Int, value: T) {  }
5.  }
```

對於以下實例，無法將一個 Array<Int> 陣列複製到 Array<Any> 陣列。Array<T> 在 T 上是不型變的。

```
1.  // copy() 函數
2.  fun copy(from: Array<Any>, to: Array<Any>) {
3.      assert(from.size == to.size)
4.      for (i in from.indices)
5.          to[i] = from[i]
6.  }
7.  fun copyValue() {
8.      val ints: Array<Int> = arrayOf(1, 2, 3)
9.      val any = Array<Any>(3) { "" }
10.     copy(ints, any)
11.     println(any.asList())
12.     // 其類型為 Array<Int>，但此處期望是 Array<Any>
13. }
```

我們唯一要確保的是，copy() 不會做任何壞事。如果想阻止它寫到 from，可以用這樣的敘述：

```
fun copy(from: Array<out Any>, to: Array<Any>) { …… }
```

這就是類型投影。我們所說的 from 不僅是一個陣列，還是一個受限制的（投影的）陣列，其只可以呼叫傳回類型為類型參數 T 的方法，如上，這表示只能呼叫 get()。同理，也可以用 in 投影一個類型。可以傳遞一個 CharSequence 陣列或一個 Object 陣列給 fill() 函數。

```
1.  // 類型投影
2.  fun fill(dest: Array<in String>, value: String) {
3.      for(i in dest.indices){
4.          dest[i] = value
5.      }
6.  }
7.  // 定義一個陣列，元素類型是 Any
8.  val arrayNulls = arrayOfNulls<Any>(2)
9.  // 將 "hello" 字串複製到 arrayNulls 陣列
10. fill(arrayNulls, "hello")
11. println(arrayNulls.asList())
```

3.2.4 委派

委派模式已經被證明是實現繼承的很好的替代方式。DelagateDerived 類別可以透過將其所有公有成員委派給指定物件來實現一個介面 Delagate。DelagateDerived 的超類型列表中的 by 子句表示 b 將在 DelagateDerived 內部儲存，並且編譯器將產生轉發給 b 的所有 Delagate 的方法。編譯器會使用 override 覆蓋的實現而非委派物件中的實現。以這種方式重新定義的成員不會在委派物件的成員中呼叫，委派物件的內部成員只能存取自己實現的介面成員。

```
1.   // 定義介面 Delagate
2.   interface Delagate {
3.       val message: String
4.       fun print()
5.   }
6.   // 定義介面 Delagate 的實現類別
7.   class DelagateImpl(val x: Int) : Delagate {
8.       override val message = "BaseImpl: x = $x"
9.       override fun print() { println(message) }
10.  }
11.  // 委派類別
12.  class DelagateDerived(b: Delagate) : Delagate by b {
13.      // 在 b 的 print 實現中不會存取到這個屬性
14.      override val message = "Message of Derived"
15.  }
16.  fun main() {
17.      val b = DelagateImpl(10)
18.      val derived = DelagateDerived(b)
19.      // 列印：BaseImpl: x = 10
20.      derived.print()
21.      // 列印：Message of Derived
22.      println(derived.message)
23.  }
```

Kotlin 支援委派屬性、延遲屬性、可觀察屬性等。語法是：val/var ＜屬性名稱＞: ＜類型＞ by ＜運算式＞。by 後面的運算式是委派類別，這是因為屬性對應的 get()（與 set()）會被委派給它的 getValue() 與 setValue() 方法。屬性的委派不必實現任何介面，但是需要提供一個 getValue() 函數（與 setValue()──對 var 屬性來說）。例如：

```
1.   // Example 類別
2.   class Example {
3.       // 委派屬性
4.       var p: String by Delegate()
```

```
5.  }
6.  // 代理類別
7.  class Delegate {
8.      operator fun getValue(thisRef: Any?, property: KProperty<*>): String {
9.          return "$thisRef, thank you for delegating '${property.name}' to me!"
10.     }
11.
12.     operator fun setValue(thisRef: Any?, property: KProperty<*>, value:
    String) {
13.         println("$value has been assigned to '${property.name}' in
    $thisRef.")
14.     }
15. }
16. val e = Example()
17. // 列印：//io.kang.chapter03.basicsyntax.Example@73a8dfcc, thank you for
    delegating '//p' to me!
18. println(e.p)
19. // 列印：//NEW has been assigned to 'p' in io.kang.chapter03.basicsyntax.
    Example@73a8 //dfcc.
20. e.p = "NEW"
```

3.3 函數和 Lambda 運算式

本節介紹函數的定義、參數、傳回類型和 Lambda 運算式的操作。

3.3.1 函數

Kotlin 用 fun 關鍵字宣告函數，函數的參數必須有顯性類型，函數參數可以有預設值，當省略對應的參數時使用預設參數值。覆蓋方法總是使用與基礎類別類型方法相同的預設參數值。當覆蓋一個帶有預設參數值的方法時，必須從簽名中省略預設參數值。如果一個預設參數在一個無預設值的參數之前，那麼該預設值只能透過使用具名引數呼叫該函數來使用。如果在預設參數之後的最後一個參數是 Lambda 運算式，那麼它作為具名引數既可以在括號內傳入，也可以在括號外傳入。函數的參數（通常是最後一個）可以用 vararg 修飾符號修飾。

如果一個函數不傳回任何有用的值，那麼它的傳回類型是 Unit。Unit 傳回型態宣告也是可選的。當函數傳回單一運算式時，可以省略大括號並且在 "=" 之後指定程式體。當傳回數值型態可由編譯器推斷時，顯性宣告傳回類型是可選的。

標有 infix 關鍵字的函數也可以使用中綴標記法（忽略該函數的點與小括號）呼叫。中綴函數必須滿足以下要求。

- 它們必須是成員函數或擴充函數。
- 它們只能有一個參數。
- 其參數不得接收可變數量的參數且不能有預設值。
- 中綴函數的優先順序低於算術運算符號、類型轉換及 rangeTo 運算符號。

```
1.  // 定義類別 A
2.  open class A {
3.      open fun foo(i: Int = 10) { /*...*/ }
4.  }
5.  // B 繼承自 A
6.  class B : A() {
7.      override fun foo(i: Int) { /*...*/ }   // 不能有預設值
8.  }
9.  // 函數 foo
10. fun foo(bar: Int = 0, baz: Int) { /*...*/ }
11. // 函數 foo
12. fun foo(bar: Int = 0, baz: Int = 1, qux: () -> Unit): Unit { /*...*/ }
13.
14. // 單運算式函數
15. fun double(x: Int) = x * 2
```

```
16. // 泛型函數
17. fun <T> asList(vararg ts: T): List<T> {
18.     val result = ArrayList<T>()
19.     for (t in ts) // ts 是一個陣列
20.         result.add(t)
21.     return result
22. }
23.
24. fun main() {
25.     foo(baz = 1) // 使用預設值 bar = 0
26.
27.     foo(1) { println("hello") }      // 使用預設值 baz = 1 的函數 foo
28.     // 使用兩個預設值 bar = 0、baz = 1 的函數 foo
29.     foo(qux = { println("hello") })
30.     foo { println("hello") }
31.     val a = arrayOf(1, 2, 3)
32.     // 可以用 * 將陣列展開
33.     val list = asList(-1, 0, *a, 4)
```

Kotlin 中的函數可以在局部作用域中宣告，作為成員函數、頂層函數及擴充函數。局部函數，即一個函數在另一個函數內部，局部函數可以存取外部函數（即閉包）的區域變數。

```
1.  data class Vertex(val neighbors: List<Vertex>)
2.  data class Graph(val vertices: List<Vertex>)
3.  fun dfs(graph: Graph) {
4.      val visited = HashSet<Vertex>()
5.      // dfs 是局部函數
6.      fun dfs(current: Vertex) {
7.          if (!visited.add(current)) return
8.          for (v in current.neighbors)
9.              dfs(v)
10.     }
11.     dfs(graph.vertices[0])
12. }
```

成員函數是在類別或物件內部定義的函數。函數可以有泛型參數,透過在函數名稱前使用中括號指定。Kotlin 支援一種稱為尾遞迴的函數式程式設計風格,允許將一些通常用循環寫的演算法改用遞迴函數來寫,進一步避免堆疊溢位的風險。當一個函數用 tailrec 修飾符號修飾並滿足所需的形式時,編譯器會最佳化該遞迴程式,留下一個快速而高效的以循環為基礎的版本。

```
1.  val eps = 1E-10
2.  // 尾遞迴函數
3.  tailrec fun findFixPoint(x: Double = 1.0): Double
4.          = if (Math.abs(x - Math.cos(x)) < eps) x else findFixPoint(Math.
    cos(x))
```

內聯函數可以加強執行時期效率,用 inline 修飾。inline 修飾符號影響函數本身和傳給它的 Lambda 運算式,所有這些都將內聯到呼叫處。

```
inline fun <T> lock(lock: Lock, body: () -> T): T { ... }
```

3.3.2 Lambda 運算式

Kotlin 中的函數是一等公民,這表示它們可以儲存在變數與資料結構中、作為參數傳遞給其他高階函數及從其他高階函數傳回,可以像操作任何其他非函數值一樣操作函數。

為促成這一點,作為靜態類型程式語言的 Kotlin 使用一系列函數類型表示函數並提供一組特定的語言結構,舉例來說,Lambda 運算式。

高階函數是將函數用作參數或傳回值的函數。fold 函數接收一個初始累積值與一個接合函數,並透過將目前累積值與每個集合元素連續接合起來代入累積值來建置傳回值。為了呼叫 fold,需要傳給它一個函數類型的實例作為參數,而在高階函數呼叫處,廣泛使用 Lambda 運算式。

```
1.  fun main() {
2.      val items = listOf(1, 2, 3, 4, 5)
3.      // Lambda 運算式是大括號括起來的程式區塊
4.      items.fold(0, {
5.          // 如果一個 Lambda 運算式有參數，前面是參數，後跟 "->"
6.          acc: Int, i: Int ->
7.          print("acc = $acc, i = $i, ")
8.          val result = acc + i
9.          println("result = $result")
10.         // Lambda 運算式中的最後一個運算式是傳回值
11.         result
12.     })
13.     // Lambda 運算式的參數類型是可選的，如果能夠將其推斷出來的話，不需要宣
    告參數類型
14.     val joinedToString = items.fold("Elements:", { acc, i -> acc + " " + i })
15.     // 列印：Elements: 1 2 3 4 5
16.     println(joinedToString)
17.     // 函數參考也可以用於高階函數呼叫
18.     val product = items.fold(1, Int::times)
19.     // 列印：120
20.     println(product)
21. }
```

Kotlin 使用類似 (Int) -> String 的一系列函數類型來處理函數的宣告。這些類型具有與函數名稱相對應的特殊標記法，即它們的參數和傳回值。

有幾種方法可以獲得函數類型的實例。

使用函數字面額的程式區塊，採用以下形式之一。

- Lambda 運算式：{ a, b -> a + b }。
- 匿名函數：fun(s: String): Int { return s.toIntOrNull() ?: 0 }。

使用已有宣告的可呼叫參考。

- 頂層、局部、成員、擴充函數：::isOdd、String::toInt。
- 頂層、成員、擴充屬性：List<Int>::size。
- 建構函數：::Regex。

函數類型的值可以透過其 invoke() 運算符號呼叫：f.invoke(x) 或直接用 f(x)。

```
1.  val stringPlus: (String, String) -> String = String::plus
2.  val intPlus: Int.(Int) -> Int = Int::plus
3.  // 列印:<-->
4.  println(stringPlus.invoke("<-", "->"))
5.  // 列印:Hello, world!
6.  println(stringPlus("Hello, ", "world!"))
7.  // 列印:2
8.  println(intPlus.invoke(1, 1))
9.  // 列印:3
10. println(intPlus(1, 2))
11. // 列印:5
12. println(2.intPlus(3))        // 類別擴充呼叫
```

一個 Lambda 運算式只有一個參數是很常見的。如果編譯器自己可以識別出簽名，也可以不用宣告唯一的參數並忽略 ->。該參數會被隱式宣告為 it：

```
ints.filter { it > 0 }          // 這個字面額是 "(it: Int) -> Boolean" 類型的
```

如果 Lambda 運算式的參數未使用，那麼可以用底線取代其名稱：

```
map.forEach { _, value -> println("$value!") }
```

3.4 集合

本節介紹 Kotlin 中的 List、Set、Map 集合及集合的操作，包含增加元素、尋找元素、過濾元素、轉換等。

3.4.1 集合概述

Kotlin 標準函數庫提供了一整套用於管理集合的工具，集合通常包含相同類型的一些（數目也可以為零）物件。集合中的物件稱為元素或項目。以下是 Kotlin 相關的集合類型。

- List 是一個有序集合，可透過索引（反映元素位置的整數）存取元素。元素可以在 List 中出現多次。List 有一組字，這些字的順序很重要並且字可以重複。

- Set 是唯一元素的集合。它反映了集合的數學抽象：一組不重複的物件。一般來説，Set 中元素的順序並不重要。

- Map（或字典）是一組鍵值對。鍵是唯一的，每個鍵都剛好對映到一個值，值可以重複。Map 對儲存物件之間的邏輯連接非常有用。

唯讀集合類型是型變的。這表示，如果類別 Rectangle 繼承自 Shape，則可以在需要 List <Shape> 的任何地方使用 List<Rectangle>。換句話説，集合類型與元素類型具有相同的類別繼承關係。Map 在值（value）類型上是型變的，但在鍵（key）類型上不是。反之，可變集合不是型變的；否則將導致執行時期故障。

如圖 3.1 所示的是 Kotlin 集合和介面的繼承關係。

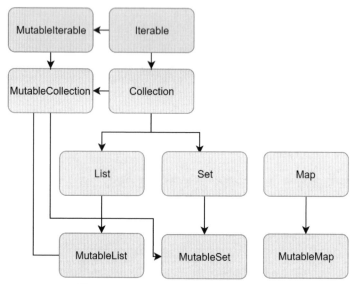

圖 3.1 Kotlin 集合和介面的繼承關係

Collection<T> 是集合層次結構的根。此介面表示一個唯讀集合的共同行
為：檢索大小、檢測是否為成員等。Collection 繼承自 Iterable<T> 介面，
它定義了反覆運算元素的操作。可以使用 Collection 作為適用於不同集合
類型的函數的參數。MutableCollection 是一個具有寫入操作的 Collection
介面，例如 add 及 remove。

List<T> 以指定的循序儲存元素，並提供使用索引存取元素的方法。索引
從 0 開始，直到最後一個元素的索引，即（list.size - 1）。List 中的元素
（包含空值）可以重複：List 可以包含任意數量的相同物件或單一物件。
如果兩個 List 在相同的位置具有相同大小和相同結構的元素，則認為它們
是相等的。MutableList 是可以進行寫入操作的 List，舉例來說，用於在特
定位置增加或刪除元素。在某些方面，List 與陣列（Array）十分類似。但
是，它們有一個重要的區別──陣列的大小是在初始化時定義的，永遠不
會改變；而 List 未預先定義大小。作為寫入操作的結果，可以更改 List 的
大小──增加、更新或刪除元素。

Set<T> 儲存唯一的元素；元素的順序通常未定義。null 元素也是唯一的，一個 Set 中只能包含一個 null。當兩個 Set 具有相同的大小並且一個 Set 中的每個元素都能在另一個 Set 中存在相同元素，則兩個 Set 相等。MutableSet 是一個帶有來自 MutableCollection 的寫入操作介面的 Set。Set 的預設實現是 LinkedHashSet，保持元素插入時的順序。另一種實現方式是 HashSet，不宣告元素的順序。

Map<K, V> 不是 Collection 介面的繼承者，但它也是 Kotlin 的一種集合類型。Map 儲存鍵值對（或項目）；鍵是唯一的，但是不同的鍵可以與相同的值配對。Map 介面提供特定的函數進行透過鍵存設定值、搜尋鍵和值等操作。無論鍵值對的順序如何，包含相同鍵值對的兩個 Map 是相等的。MutableMap 是一個具有寫入操作的 Map 介面，可以使用該介面增加一個新的鍵值對或更新指定鍵的值。Map 的預設實現是 LinkedHashMap，反覆運算 Map 時保持元素插入時的順序。反之，另一種實現方式是 HashMap，不宣告元素的順序。

```
1.  val bob = Person("Bob", 31)
2.  val people = listOf<Person>(Person("Adam", 20), bob, bob)
3.  val people2 = listOf<Person>(Person("Adam", 20), Person("Bob", 31), bob)
4.  // 列印：false
5.  println(people == people2)
6.  bob.age = 32
7.  // 列印：false
8.  println(people == people2)
9.  // 可變 List
10. val numbers = mutableListOf(1, 2, 3, 4)
11. numbers.add(5)
12. numbers.removeAt(1)
13. numbers[0] = 0
14. // 打亂元素順序
15. numbers.shuffle()
16. // 列印：[4, 5, 0, 3]
```

```
17. println(numbers)
18. // Set
19. val numbersBackwards = setOf(4, 3, 2, 1)
20. println("The sets are equal: ${numbers == numbersBackwards}")
21.
22. // Map
23. val numbersMap = mapOf("key1" to 1, "key2" to 2, "key3" to 3, "key4" to 1)
24. // 列印：All keys: [key1, key2, key3, key4]
25. println("All keys: ${numbersMap.keys}")
26. // 列印：All values: [1, 2, 3, 1]
27. println("All values: ${numbersMap.values}")
28.
29. val anotherMap = mapOf("key2" to 2, "key1" to 1, "key4" to 1, "key3" to 3)
30. // 列印：The maps are equal: true
31. println("The maps are equal: ${numbersMap == anotherMap}")
32. // 可變 Map
33. val mutableNumbersMap = mutableMapOf("one" to 1, "two" to 2)
34. mutableNumbersMap.put("three", 3)
35. mutableNumbersMap["one"] = 11
```

建立集合的最常用方法是使用標準函數庫函數 listOf<T>()、setOf<T>()、
mutableListOf<T>()、mutableSetOf<T>()。如果以逗點分隔的集合元素清單
作為參數，編譯器會自動檢測元素類型。建立空集合時，必須明確指定集
合類型。

同樣地，Map 也有這樣的函數——mapOf() 與 mutableMapOf()。對映的鍵
和值作為 Pair 物件傳遞（通常使用中綴函數 to 建立）。可以建立寫入 Map
並使用寫入操作填充它。apply() 函數在初始化時使用。

建立沒有任何元素的集合的函數有 emptyList()、emptySet() 與 emptyMap()。
建立空集合時，應指定集合將包含的元素類型。對於 List，有一個接收
List 的大小與初始化函數的建構函數，該初始化函數根據索引定義元素的
值。

```
1.  // 初始化集合
2.  val numbersSet = setOf("one", "two", "three", "four")
3.  // 初始化空集合
4.  val emptySet = mutableSetOf<String>()
5.  // 初始化 Map
6.  val numbersMap1 = mapOf("key1" to 1, "key2" to 2, "key3" to 3, "key4" to 1)
7.  // 初始化可變 Map
8.  val numbersMap2 = mutableMapOf<String, String>().apply { this["one"] =
    "1"; this["two"] = "2" }
9.  // 初始化空 List
10. val empty = emptyList<String>()
11. val doubled = List(3, { it * 2 })
12. // 列印：[0, 2, 4]
13. println(doubled)
```

要建立與現有集合具有相同元素的集合，可以使用複製操作。標準函數庫
中的集合複製操作建立了具有相同元素參考的淺複製集合。因此，對集合
元素所做的更改會反映在其所有備份中。但是 toList()、toMutableList()、
toSet() 等建立了一個具有相同元素的新集合，如果在來源集合中增加或刪
除元素，則不會影響備份。備份也可以獨立於來源集合進行更改。這些函
數還可用於將集合轉為其他類型，舉例來說，根據 List 建置 Set，反之亦
然。

```
1.  val sourceList = mutableListOf(1, 2, 3)
2.  // 複製元素到新 List
3.  val copyList = sourceList.toMutableList()
4.  sourceList.add(4)
5.  // 列印：Copy size: 3
6.  println("Copy size: ${copyList.size}")
```

Kotlin 支援使用反覆運算器檢查集合中的元素，反覆運算器可以在不曝露
集合內部結構的情況下，順序檢查集合中的元素。反覆運算器適用於一
個一個處理集合元素的場景。Set 和 List 的 iterator() 函數提供了反覆運算

器,反覆運算器初始指向集合的第一個元素,next() 函數傳回目前元素並指向下一個元素。反覆運算器可以檢查集合中的元素,但不能取得元素。如果需要再次檢查集合,需要重新建立一個反覆運算器。此外,用 for 或 forEach 敘述也可以檢查集合中的元素。

```
1. val numberStrs = listOf("one", "two", "three", "four")
2. // 檢查集合元素
3. for (item in numberStrs) {
4.     println(item)
5. }
6. numberStrs.forEach {
7.     println(it)
8. }
```

List 有一個特殊的反覆運算器 ListIterator,它支援雙向檢查,後續檢查用 hasPrevious()、previous(),可以用 nextIndex() 和 previousIndex() 取得元素索引。

可變集合提供 MutableIterator,當檢查集合時,可以用 remove 移除元素。MutableListIterator 可以修改、增加元素。

```
1.  val numberStrs1 = mutableListOf("one", "two", "three", "four")
2.  val mutableIterator = numberStrs1.iterator()
3.  // 使用反覆運算器操作集合元素
4.  mutableIterator.next()
5.  mutableIterator.remove()
6.  // 列印:After removal: [two, three, four]
7.  println("After removal: $numberStrs1")
8.  val mutableListIterator = numberStrs1.listIterator()
9.  mutableListIterator.next()
10. mutableListIterator.add("two")
11. mutableListIterator.next()
12. mutableListIterator.set("three")
13. // 列印:[two, two, three, four]
14. println(numberStrs1)
```

Kotlin 可以使用 rangeTo 建立一連串數值，通常用 .. 替代 rangeTo。這些數值可以用 for 敘述檢查。如果要進行後續檢查，可以用 downTo。這些數值的間隔通常是 1，也可以用 step 指定步進值。如果不需要檢查最後一個元素，可以用 until。可以用 in、!in 判斷某個元素是否在區間內。

```
1.  // 列印 1 3 5 7 9，不包含 10
2.  for (i in 1 until 10 step 2) {
3.      print(i)
4.
5.  val rangeInt = 1..10 step 2
6.  // 列印：true
7.  println(3 in rangeInt)
8.  // 列印：true
9.  println(2 !in rangeInt)
10. }
```

Kotlin 提供序列類型 Sequence<T>。當反覆運算過程包含多個步驟時，每一步執行完，會產生一個結果，下面的步驟會在這個結果的基礎上執行。序列中的每一個元素都要依次執行操作步驟，反覆運算器對集合中的所有元素執行完一個操作過程後再執行下一個。序列可避免產生中間結果，可以加強整個處理鏈的效能。

通常可以用 sequenceOf() 建置序列；可以用 asSequence() 將 List、Set 轉為序列；可以用 generateSequence() 顯示指定序列的第一個元素，函數傳回 null 時序列停止增長；序列也可以用群組塊產生，函數包含 Lambda 運算式，包含 yield()、yieldAll() 函數。yield() 的參數是一個元素，yieldAll() 的參數可以是一個集合，也可以是一個序列，這個序列可以無限長。

```
1.  // 用函數建置序列
2.  val oddNumbers = generateSequence(1) { it + 2 }
3.  // 列印：[1, 3, 5, 7, 9]
4.  println(oddNumbers.take(5).toList())
5.
```

```
6.  // 限長序列
7.  val oddNumbersLessThan10 = generateSequence(1) { if (it < 10) it + 2 else
    null }
8.  // 列印：6
9.  println(oddNumbersLessThan10.count())
10. val oddNumbers1 = sequence {
11.     yield(1)
12.     yieldAll(listOf(3, 5))
13.     yieldAll(generateSequence(7) { it + 2 })
14. }
15. // 列印：[1, 3, 5, 7, 9]
16. println(oddNumbers1.take(5).toList())
```

序列支援無狀態的操作，如 filter、take、map、drop 等。下面的實例展示了序列的處理過程，對每個元素都會執行 filter、map、take 方法，在找出4 個元素後，不再檢查其餘元素。

```
1.  val words = "The quick brown fox jumps over the lazy dog".split(" ")
2.  // 將集合轉為序列
3.  val wordsSequence = words.asSequence()
4.  val lengthsSequence = wordsSequence
5.          .filter { println("filter: $it"); it.length > 3 }
6.          .map { println("length: ${it.length}"); it.length }
7.          .take(4)
8.  // 列印
9.  // filter: The
10. // filter: quick
11. // length: 5
12. // filter: brown
13. // length: 5
14. // filter: fox
15. // filter: jumps
16. // length: 5
17. // filter: over
18. // length: 4
```

```
19. // [5, 5, 5, 4]
20. println(lengthsSequence.toList())
```

3.4.2 集合操作

Kotlin 標準函數庫提供了很多函數操作集合，如 set、add、search、sorting、filtering、transformations 等。

集合有以下操作：轉換（transformations）、過濾（filtering）、加減（plus and minus operators）、分組（grouping）、取得部分集合（retrieving collection parts）、排序（ordering）、聚合操作（aggregate operations）。

集合操作不會改變集合本身，而是產生一個新的集合儲存處理後的結果。此外，還可以指定一個可變物件儲存處理後的結果。

```
1.  val numbers = listOf("one", "two", "three", "four")
2.  val filterResults = mutableListOf<String>()
3.  // 對 numbers 進行過濾操作，複製到新集合
4.  numbers.filterTo(filterResults) { it.length > 3 }
5.  // 列印：[three, four]
6.  println(filterResults)
```

Kotlin 提供了很多集合轉換的擴充操作，這些操作會產生新的集合。下面介紹幾種轉換操作：對映（mapping），用 Lambda 運算式對目前集合中的元素進行處理，產生一個新的集合，兩個集合中元素的順序一致。當產生 null 元素時，可以用 mapNotNull() 函數將其過濾掉。當轉換對映時，可以用 mapKeys 處理 key 值，或用 mapValues 處理 value 值。

```
1.  val numbers1 = setOf(1, 2, 3)
2.  // 集合對映操作
3.  println(numbers1.map { it * 3 })
4.  println(numbers1.mapIndexed { idx, value -> value * idx })
5.  // 用 mapNotNull 函數過濾 null 元素
```

```
6.  println(numbers1.mapNotNull { if ( it == 2) null else it * 3 })
7.  println(numbers1.mapIndexedNotNull { idx, value -> if (idx == 0) null
    else value * idx })
8.
9.  val numbersMap = mapOf("key1" to 1, "key2" to 2, "key3" to 3, "key11" to 11)
10. // 列印:{KEY1=1, KEY2=2, KEY3=3, KEY11=11}
11. println(numbersMap.mapKeys { it.key.toUpperCase() })
12. // 列印:{key1=5, key2=6, key3=7, key11=16}
13. println(numbersMap.mapValues { it.value + it.key.length })
```

雙路合併（zip）操作，將兩個集合中的元素合併為一個 List，List 中每個元素成對出現，是 Pair 類型的。如果兩個集合大小不一樣，zip 操作取較小的集合長度。此外，在 zip 操作的基礎上，還可以進行 Lambda 操作，最後傳回的 List 元素集合就是由 Lambda 操作後的元素組成的。相反，使用 unzip 函數可以進行反向操作。

```
1.  val colors = listOf("red", "brown", "grey")
2.  val animals = listOf("fox", "bear", "wolf")
3.  // 使用 zip 合併集合
4.  // 列印:[(red, fox), (brown, bear), (grey, wolf)]
5.  println(colors zip animals)
6.
7.  val twoAnimals = listOf("fox", "bear")
8.  // 列印:[(red, fox), (brown, bear)]
9.  println(colors.zip(twoAnimals))
10. // 列印:[The Fox is red, The Bear is brown, The Wolf is grey]
11. println(colors.zip(animals) { color, animal -> "The ${animal.capitalize()}
    is $color"})
12. val numberPairs = listOf("one" to 1, "two" to 2, "three" to 3, "four" to 4)
13. // 列印:([one, two, three, four], [1, 2, 3, 4])
14. println(numberPairs.unzip())
```

連結（association）操作，可以對集合中的元素進行處理，以產生一個對映。基本的關聯函數是 associateWith()，集合中的元素作為對映的 key，轉換後的元素作為對映的 value。associateBy() 函數將集合中的元素作為對映的 value，轉換後的元素作為對映的 key。

```
1.  val numbers2 = listOf("one", "two", "two", "four")
2.  // 集合連結，列印：{one=3, two=3, four=4}
3.  println(numbers2.associateWith { it.length })
4.  // 列印：{O=one, T=two, F=four}
5.  println(numbers2.associateBy { it.first().toUpperCase() })
6.  // 列印：{O=3, T=3, F=4}
7.  println(numbers2.associateBy(keySelector = { it.first().toUpperCase() },
    valueTransform = { it.length }))
```

打平（flattern）操作，flattern() 函數將許多集合中的元素放在一個集合中；flatternMap() 函數對許多集合進行對映操作，然後再打平。

```
1.  val numberSets = listOf(setOf(1, 2, 3), setOf(4, 5, 6), setOf(1, 2))
2.  // 使用 flattern 打平集合，列印：[1, 2, 3, 4, 5, 6, 1, 2]
3.  println(numberSets.flatten())
```

字串處理操作，可以用 joinToString() 函數將集合中的元素產生一個字串，用 joinTo() 函數將產生的字串連接在另一個字串後。

```
1.  val numbers3 = listOf("one", "two", "three", "four")
2.  // 列印：one, two, three, four
3.  println(numbers3.joinToString())
4.  val listString = StringBuffer("The list of numbers: ")
5.  numbers3.joinTo(listString)
6.  // 列印：The list of numbers: one, two, three, four
7.  println(listString)
```

過濾操作在集合處理中使用廣泛。Kotlin 使用述詞函數實現過濾操作，使用 Lambda 運算式處理集合的元素，並傳回一個布林值。如果需要使用集合元素的索引，可以使用 filterIndexed() 函數。如果需要使用相反條件過濾，可以使用 filterNot() 函數。如果需要過濾類型，可以使用 filterIsInstance<T>() 函數。需要過濾空，可以使用 filterNotNull() 函數。

```
1.  val numbersMap1 = mapOf("key1" to 1, "key2" to 2, "key3" to 3, "key11" to 11)
2.  val filteredMap = numbersMap1.filter { (key, value) -> key.endsWith("1")
    && value > 10}
3.  // 列印：{key11=11}
4.  println(filteredMap)
5.  val numbers4 = listOf("one", "two", "three", "four")
6.  val filteredIdx = numbers4.filterIndexed { index, s -> (index != 0) &&
    (s.length < 5) }
7.  val filteredNot = numbers4.filterNot { it.length <= 3 }
8.  // 列印：[two, four]
9.  println(filteredIdx)
10. // 列印：[three, four]
11. println(filteredNot)
```

可以使用 partition() 函數對集合進行劃分,一部分滿足過濾條件,另一部分不滿足過濾條件。可以使用 any()(至少有一個)、none()(一個也沒有)、all()(全部滿足)來檢驗述詞函數。

集合還可以進行加減操作,運算子的左值是集合,右值可以是一個元素或集合。

```
1.  val numbers5 = listOf("one", "two", "three", "four")
2.
3.  val plusList = numbers5 + "five"
4.  val minusList = numbers5 - listOf("three", "four")
5.  // 列印：[one, two, three, four, five]
6.  println(plusList)
7.  // 列印：[one, two]
8.  println(minusList)
```

Kotlin 還提供了分組操作。groupBy 函數使用一個 Lambda 運算式,傳回一個對映。對映的 key 是 Lambda 運算式傳回的值,value 是集合中的元素。

```
1.  val numbers6 = listOf("one", "two", "three", "four", "five")
2.  // 列印：{O=[one], T=[two, three], F=[four, five]}
```

```
3.  println(numbers6.groupBy { it.first().toUpperCase() })
4.  // 列印：{o=[ONE], t=[TWO, THREE], f=[FOUR, FIVE]}
5.  println(numbers6.groupBy(keySelector = { it.first() }, valueTransform = {
    it.toUpperCase() }))
```

Kotlin 提供了豐富的操作以取得集合的一部分。slice() 函數根據指定索引
傳回集合中的元素。take() 函數從第一個元素開始，截取指定個數的集合
元素；如果截取個數大於集合長度，傳回整個集合元素。drop() 函數從
第一個元素開始，捨棄指定個數的元素，然後傳回其餘元素。如果要使
用述詞函數，可以用 takeWhile()、takeLastWhile()、dropWhile() 函數、
dropLastWhile()。chunked() 函數將集合分段，檢查集合元素，達到指定數
量，產生一個 List，直到最後一個元素。windowed() 函數從第一個元素開
始進行檢查，可以指定視窗大小。此外，windowed() 函數可以指定步進值
step 等參數。如果滑動視窗中只有兩個元素，可以用 zipWithNext() 函數。

```
1.  val numbers7 = listOf("one", "two", "three", "four", "five", "six")
2.  // 列印：[two, three, four]
3.  println(numbers7.slice(1..3))
4.  // 列印：[one, three, five]
5.  println(numbers7.slice(0..4 step 2))
6.  // 列印：[four, six, one]
7.  println(numbers7.slice(setOf(3, 5, 0)))
8.  val numbers8 = listOf("one", "two", "three", "four", "five", "six")
9.  // 列印：[one, two, three]
10. println(numbers8.take(3))
11. // 列印：[four, five, six]
12. println(numbers8.takeLast(3))
13. // 列印：[two, three, four, five, six]
14. println(numbers8.drop(1))
15. // 列印：[one]
16. println(numbers8.dropLast(5))
17. // 列印：[one, two, three]
18. println(numbers8.takeWhile { !it.startsWith('f') })
```

```
19. // 列印：[four, five, six]
20. println(numbers8.takeLastWhile { it != "three" })
21. // 列印：[three, four, five, six]
22. println(numbers8.dropWhile { it.length == 3 })
23. // 列印：[one, two, three, four]
24. println(numbers8.dropLastWhile { it.contains('i') })
25. val numbers9 = (0..13).toList()
26. // 列印：[[0, 1, 2], [3, 4, 5], [6, 7, 8], [9, 10, 11], [12, 13]]
27. println(numbers9.chunked(3))
28. // 列印：[3, 12, 21, 30, 25]
29. println(numbers9.chunked(3) { it.sum() })
30. val numbers10 = (1..10).toList()
31. // 列印：[[1, 2, 3], [2, 3, 4], [3, 4, 5], [4, 5, 6], [5, 6, 7], [6, 7,
    8], //[7, 8, 9], [8, 9, 10]]
32. println(numbers10.windowed(3))
33. // 列印：[[1, 2, 3], [3, 4, 5], [5, 6, 7], [7, 8, 9], [9, 10]]
34. println(numbers10.windowed(3, step = 2, partialWindows = true))
35. // 列印：[6, 9, 12, 15, 18, 21, 24, 27]
36. println(numbers10.windowed(3) { it.sum() })
37. // 列印：[(1, 2), (2, 3), (3, 4), (4, 5), (5, 6), (6, 7), (7, 8), (8, 9),
    (9, //10)]
38. println(numbers10.zipWithNext())
39. // 列印：[3, 5, 7, 9, 11, 13, 15, 17, 19]
40. println(numbers10.zipWithNext() { s1, s2 -> s1 + s2})
```

如果要取得集合中的單一元素，可以使用 elementAt() 按位置取得，因 first() 取得第一個元素，用 last() 取得最後一個元素，也可以在 first() 或 last() 後用述詞函數按條件取得。用 random() 函數可隨機取得元素，用 contains() 檢測是否存在某個元素，用 containsAll() 檢測是否存在多個元素，用 isEmpty()、isNotEmpty() 判斷集合是否為空。

Kotlin 可以用 Comparable 介面對自訂的類型進行排序，Kotlin 附帶的類型預設支援排序，數值型按數值大小排序，字元型按照字母順序排序。此外，還可以用 Comparator 自訂順序，compareBy() 是 Comparator 的簡單寫法。自然順序可以用 sorted() 和 sortedDescending() 進行昇冪或降冪操作。
自訂順序可以用 sortedBy() 和 sortedByDescending() 進行昇冪或降冪操作。
倒序可以用 reversed()、asReversed()。隨機順序可以用 shuffled()。

```
1.  val unSortednumbers = listOf("one", "two", "three", "four")
2.
3.  val lengthComparator = Comparator { str1: String, str2: String -> str1.
    length - str2.length }
4.  // 列印：[one, two, four, three]
5.  println(unSortednumbers.sortedWith(lengthComparator))
6.  // 列印：[one, two, four, three]
7.  println(unSortednumbers.sortedWith(compareBy { it.length }))
8.  // 列印：Sorted ascending: [four, one, three, two]
9.  println("Sorted ascending: ${unSortednumbers.sorted()}")
10. // 列印：Sorted descending: [two, three, one, four]
11. println("Sorted descending: ${unSortednumbers.sortedDescending()}")
12.
13. val sortedNumbers = unSortednumbers.sortedBy { it.length }
14. // 列印：Sorted by length ascending: [one, two, four, three]
15. println("Sorted by length ascending: $sortedNumbers")
16. val sortedByLast = unSortednumbers.sortedByDescending { it.last() }
17. // 列印：Sorted by the last letter descending: [four, two, one, three]
18. println("Sorted by the last letter descending: $sortedByLast")
19. // 列印：[four, three, two, one]
20. println(unSortednumbers.reversed())
21. // 列印：[two, three, four, one]
22. println(unSortednumbers.shuffled())
```

集合的聚合操作函數有求最小值 min()、最大值 max()、平均值 average()、
求和 sum()、計數 count()，帶有函數的求最大值 / 最小值為 maxBy()/
minBy()，Comparator 物件的求最大值 / 最小值為 maxWith()/minWith()，帶
有述詞函數的求和為 sumBy()，傳回 Double 類型的求和為 sumDouble()，
累加為 fold()、reduce()，fold() 有初值，reduce() 開始時用前兩個元素作為
參數。

```
1.  val numbersMerge = listOf(6, 42, 10, 4)
2.  // 列印：Count: 4
3.  println("Count: ${numbersMerge.count()}")
4.  // 列印：Max: 42
5.  println("Max: ${numbersMerge.max()}")
6.  // 列印：Min: 4
7.  println("Min: ${numbersMerge.min()}")
8.  // 列印：Average: 15.5
9.  println("Average: ${numbersMerge.average()}")
10. // 列印：Sum: 62
11. println("Sum: ${numbersMerge.sum()}")
12. val min3Remainder = numbersMerge.minBy { it % 3 }
13. // 列印：6
14. println(min3Remainder)
15. val max3Remainder = numbersMerge.maxWith(compareBy { it - 10 })
16. // 列印：42
17. println(max3Remainder)
18. // 列印：124
19. println(numbersMerge.sumBy { it * 2 })
20. // 列印：31.0
21. println(numbersMerge.sumByDouble { it.toDouble() / 2 })
22. val sum = numbersMerge.reduce { sum, element -> sum + element }
23. // 列印：62
24. println(sum)
25. val sumDoubled = numbersMerge.fold(0) { sum, element -> sum + element * 2
    } // 列印：124
26. println(sumDoubled)
```

集合可以透過 add() 增加單一元素，addAll() 增加多個元素，remove() 刪除
元素，retainAll() 保留符合條件的元素，clear() 清空集合，minusAsign() 和
minus() 刪除元素。

3.4.3　List、Set、Map 相關操作

List 是使用廣泛的集合，List 可以按索引取元素，getOrElse() 函數若取不到元素就傳回預設值，getOrNull() 函數若取不到元素就傳回 null。對於有序 List，可以用 binarySearch() 進行二分尋找，此外，可以自訂排序規則。

```
1.  val listNuumbers = listOf(1, 2, 3, 2)
2.  // 列印:1
3.  println(listNuumbers.get(0))
4.  // 列印:1
5.  println(listNuumbers[0])
6.  // 列印:null
7.  println(listNuumbers.getOrNull(5))            // null
8.  // 列印:5
9.  println(listNuumbers.getOrElse(5, {it}))      // 5
10. // 列印:[1, 2]
11. println(listNuumbers.subList(0, 2))
12. // 列印:1
13. println(listNuumbers.indexOf(2))
14. // 列印:3
15. println(listNuumbers.lastIndexOf(2))
16. // 列印:2
17. println(listNuumbers.indexOfFirst { it > 2})
18. // 列印:2
19. println(listNuumbers.indexOfLast { it % 2 == 1})
20.
21. listNuumbers.sorted()
22. // 列印:1
23. println(listNuumbers.binarySearch(2))         // 3
24. // 列印:-5
25. println(listNuumbers.binarySearch(4))         // -5
26. // 列印:0
27. println(listNuumbers.binarySearch(1, 0, 2))   // -3
```

Set 提供了求交集 intersect()、合併 union()、剔除交集元素 subtract() 等操作。

```
1. val setNnumbers = setOf("one", "two", "three")
2. // 列印 : [one, two, three, four, five]
3. println(setNnumbers union setOf("four", "five"))
4. // 列印 : [one, two]
5. println(setNnumbers intersect setOf("two", "one"))
6. // 列印 : [one, two]
7. println(setNnumbers subtract setOf("three", "four"))
8. // 列印 : [one, two]
9. println(setNnumbers subtract setOf("four", "three")) // same output
```

Map 提供了取鍵、值操作，過濾鍵、值操作，plus 和 minus 操作，以及增加和更新操作。

```
1.  val mapNumbers = mapOf("one" to 1, "two" to 2, "three" to 3)
2.  // 列印 : 1
3.  println(mapNumbers.get("one"))
4.  // 列印 : 1
5.  println(mapNumbers["one"])
6.  // 列印 : 10
7.  println(mapNumbers.getOrDefault("four", 10))
8.  // 列印 : null
9.  println(mapNumbers["five"])                // null
10. // 列印 : [one, two, three]
11. println(mapNumbers.keys)
12. // 列印 : [1, 2, 3]
13. println(mapNumbers.values)
14. // 列印 : {}
15. println(mapNumbers.filter { (key, value) -> key.endsWith("1") && value >
    10})
16. // 列印 : {}
17. println(mapNumbers.filterKeys { it.endsWith("1") })
18. // 列印 : {one=1, two=2, three=3}
19. println(mapNumbers.filterValues { it < 10 })
20. // 列印 : {one=1, two=2, three=3, four=4}
21. println(mapNumbers + Pair("four", 4))
```

```
22. // 列印：{one=10, two=2, three=3}
23. println(mapNumbers + Pair("one", 10))
24. // 列印：{one=11, two=2, three=3, five=5}
25. println(mapNumbers + mapOf("five" to 5, "one" to 11))
26. // 列印：{two=2, three=3}
27. println(mapNumbers - "one")
28. // 列印：{one=1, three=3}
29. println(mapNumbers - listOf("two", "four"))
```

3.5 程式碼協同

本節介紹 Kotlin 中程式碼協同的基礎概念和進階用法。

3.5.1 程式碼協同基礎

kotlinx.coroutines 是 Kotlin 的程式碼協同函數庫。下面是一個程式碼協同的實例：

```
1.  import kotlinx.coroutines.*
2.  fun main() {
3.      // 在後台啟動一個新的程式碼協同並繼續
4.      GlobalScope.launch {
5.          // 非阻塞地等待 1 秒（預設時間單位是毫秒）
6.          delay(1000L)
7.          // 在延遲後列印輸出
8.          println("World!")
9.      }
10.     // 程式碼協同已在等待時，主執行緒還在繼續
11.     println("Hello,")
12.     // 但是這個運算式阻塞了主執行緒
13.     runBlocking {
14.         // ……延遲 2 秒來保障 JVM 的存活
```

```
15.        delay(2000L)
16.    }
17. }
```

本質上,程式碼協同是輕量級的執行緒。它們在某些 coroutineScope 上下文中與 launch 程式碼協同建構元一起啟動。GlobalScope 像守護執行緒,在 GlobalScope 中啟動的活動程式碼協同並不會使處理程序保活。可以將 GlobalScope.launch{⋯} 取代為 thread{⋯},將 delay(⋯) 取代為 Thread. sleep(⋯) 來達到同樣的目的。delay 是一個特殊的暫停函數,它不會造成執行緒阻塞,但是會暫停程式碼協同,並且只能在程式碼協同中使用。可顯性地用 runBlocking 程式碼協同建構元來阻塞執行緒。可以用 join 顯性(以非阻塞方式)地等待所啟動的後台作業執行結束。可以在執行操作所在的指定作用域內啟動程式碼協同,而非像通常使用執行緒(執行緒總是全域的)那樣在 GlobalScope 中啟動。還可以使用 coroutineScope 建構元宣告自己的作用域。

runBlocking 與 coroutineScope 的主要區別在於,後者在等待所有子程式碼協同執行完畢時不會阻塞目前執行緒。如果將 launch{⋯} 內部的程式區塊分析到獨立的函數中,需要用 suspend 修飾。

❑ 取消程式碼協同的執行

程式碼協同是可以被取消的,可以用 cancel() 取消程式碼協同。程式碼協同的取消是協作的。一段程式碼協同程式必須協作才能被取消。執行下面的範例程式,可以看到它連續列印出了 "I'm sleeping",甚至在呼叫取消後,作業仍然執行了 5 次循環反覆運算並執行到結束為止。

```
1. fun main() {
2.    runBlocking {
3.        val startTime = System.currentTimeMillis()
4.        val job = launch(Dispatchers.Default) {
5.            var nextPrintTime = startTime
```

```
6.              var i = 0
7.              while (i < 5) {
8.                  // 一個執行計算任務的循環，只是為了佔用 CPU
9.                  // 每秒列印訊息兩次
10.                 if (System.currentTimeMillis() >= nextPrintTime) {
11.                     println("job: I'm sleeping ${i++} ...")
12.                     nextPrintTime += 500L
13.                 }
14.             }
15.         }
16.     // 等待一段時間
17.     delay(1300L)
18.     println("main: I'm tired of waiting!")
19.     // 取消一個作業並且等待它結束
20.     job.cancelAndJoin()
21.     println("main: Now I can quit.")
22.     }
23. }
```

計算任務是可以取消的，可以顯性檢查取消狀態。將 while(i<5) 取代為 while(isActive)，在循環到第三次後，任務就被取消了。當程式碼協同取消時，可以在 finally() 函數中釋放資源。取消一個程式碼協同的理由是它可能逾時，可以使用 withTimeout()、withTimeoutOrNull() 函數在程式碼協同逾時後取消程式碼協同。

```
1. // 列印：I'm sleeping 0 ...
2. // I'm sleeping 1 ...
3. // I'm sleeping 2 ...
4. withTimeout(1300L) {
5.     repeat(1000) { i ->
6.             println("I'm sleeping $i ...")
7.         delay(500L)
8.     }
9. }
```

❑ 組合暫停函數

如果要根據第一個函數的結果來決定是否需要呼叫第二個函數或決定如何
呼叫第二個函數，可使用普通的順序進行呼叫。下面列舉的實例的耗時是
兩個函數耗時之和。

```
1.  suspend fun doSomethingUsefulOne(): Int {
2.      delay(1000L) // 假設我們在這裡做了一些有用的事
3.      return 13
4.  }
5.
6.  suspend fun doSomethingUsefulTwo(): Int {
7.      delay(1000L) // 假設我們在這裡也做了一些有用的事
8.      return 29
9.  }
10.
11. fun main() = runBlocking<Unit> {
12.     val time = measureTimeMillis {
13.         val one = doSomethingUsefulOne()
14.         val two = doSomethingUsefulTwo()
15.         // 列印:The answer is 42
16.          println("The answer is ${one + two}")
17.     }
18.     // 列印 Completed in 2010 ms
19.     println("Completed in $time ms")
20. }
```

當然如果兩個函數沒有依賴，可以用 async 進行平行處理。async 傳回一個
Deferred，一個輕量級的非阻塞 future，可以使用 .await()（經過一定的時
間）獲得它的最後結果，但是 Deferred 也是一個作業，是可以取消的。

```
1.  val time = measureTimeMillis {
2.      val one = async { doSomethingUsefulOne() }
3.      val two = async { doSomethingUsefulTwo() }
4.      // 列印:The answer is 42
5.      println("The answer is ${one.await() + two.await()}")
```

```
6.  }
7.  // 列印：Completed in 1027 ms
8.  println("Completed in $time ms")
```

可 以 使 用 async 進 行 結 構 化 平 行 處 理 ， 如 下 例 所 示 。 如 果 在
concurrentSum() 函數內部發生了錯誤，並且拋出了一個例外，那麼所有在
作用域中啟動的程式碼協同都會被取消。取消始終透過程式碼協同的層次
結構來進行傳遞。

```
1.  suspend fun concurrentSum(): Int = coroutineScope {
2.      val one = async { doSomethingUsefulOne() }
3.      val two = async { doSomethingUsefulTwo() }
4.      one.await() + two.await()
5.  }
```

❏ 程式碼協同上下文與排程器

程式碼協同總是執行在一些以 CoroutineContext 類型為代表的上下文中，
它們被定義在 Kotlin 的標準函數庫裡。程式碼協同上下文包含一個程式碼
協同排程器，它確定了哪些執行緒或與執行緒相對應的程式碼協同執行。
程式碼協同排程器可以將程式碼協同限制在一個特定的執行緒內執行，或
將其排程到一個執行緒池，或讓其不受限地執行。

```
1.   fun main() = runBlocking<Unit> {
2.       launch {
3.           // 執行在父程式碼協同的上下文中，即 runBlocking 主程式碼協同
4.           println("main runBlocking      : I'm working in thread ${Thread.
    currentThread().name}")
5.       }
6.       launch(Dispatchers.Unconfined) {
7.           // 不受限的——將工作在主執行緒中
8.           println("Unconfined            : I'm working in thread ${Thread.
    currentThread().name}")
9.       }
10.      launch(Dispatchers.Default) {
```

```
11.          // 將取得預設排程器
12.          println("Default                    : I'm working in thread ${Thread.
    currentThread().name}")
13.     }
14.     launch(newSingleThreadContext("MyOwnThread")) {
15.          // 將獲得一個新的執行緒
16.          println("newSingleThreadContext: I'm working in thread ${Thread.
    currentThread().name}")
17.     }
18. }
```

當呼叫 launch {…} 時不傳參數，它會從啟動了它的 CoroutineScope 中繼承上下文（及排程器）。在上述實例中，它從 main 執行緒中的 runBlocking 主程式碼協同繼承了上下文。

Dispatchers.Unconfined 是一個特殊的排程器且似乎也執行在 main 執行緒中，但實際上，它是一種不同的機制。

當程式碼協同在 GlobalScope 中啟動的時候使用，代表 Dispatchers.Default 使用了共用的後台執行緒池，所以 GlobalScope.launch {…} 也可以使用相同的排程器 launch(Dispatchers. Default) {…}。

newSingleThreadContext 為程式碼協同的執行啟動了一個執行緒。專用線程是一種非常昂貴的資源。在真實的應用程式中程式碼協同和執行緒都必須被釋放，當不再需要的時候，使用 close() 函數釋放，或將 newSingleThreadContext 儲存在一個頂層變數中使其在整個應用程式中被重用。

3.5.2 程式碼協同進階

Kotlin 的流（Flow）可以處理暫停函數非同步傳回的多個值。以下面的範例就定義了一個流，這個流不阻塞主執行緒，用 emit() 函數發射流，用 collect() 函數收集流。flow{} 建置區塊的程式可以暫停，函數 foo() 不再有 suspend 識別符號。流採用與程式碼協同同樣的協作取消方式 —— withTimeoutOrNull，當逾時時取消並停止執行其程式。

```
1.  fun foo(): Flow<Int> = flow {
2.      // 流建構元
3.      for (i in 1..3) {
4.          delay(100) // 假設我們在這裡做了一些有用的事情
5.          emit(i) // 發送下一個值
6.      }
7.  }
8.  fun main() = runBlocking<Unit> {
9.      // 啟動平行處理的程式碼協同以驗證主執行緒並未阻塞
10.     launch {
11.         for (k in 1..3) {
12.             println("I'm not blocked $k")
13.             delay(100)
14.         }
15.     }
16.     // 收集這個流
17.     foo().collect { value -> println(value) }
18. }
```

可以用 flowOf() 定義一個發射固定值集的流，用 .asFlow() 將各種集合與序列轉為流。流的過渡運算符號，如 map、filter，應用於上游流，傳回下游流。流的轉換運算符號，常用的如 transform，可以實施更複雜的轉換。限長過渡運算符號，如 take，在流觸及對應限制的時候會將其取消。

```
1.  suspend fun performRequest(request: Int): String {
2.      delay(1000) // 模仿長時間執行的非同步工作
3.      return "response $request"
4.  }
5.  fun main() = runBlocking<Unit> {
6.      (1..3).asFlow() // 一個請求流
```

```
7.            .map { request -> performRequest(request) }
8.            .collect { response -> println(response) }
9. }
```

流的末端運算符號有 collect、toList 和 toSet，它們可將流轉為集合，first()
取得第一個值，single() 確保流發射單一值，reduce()、fold() 將流歸約到單
一值。

可以在流上使用 buffer 運算符號來平行處理執行發射元素的程式及收集的
程式，而非順序執行它們。

```
1.  fun log(msg: String) = println("[${Thread.currentThread().name}] $msg")
2.  fun foo(): Flow<Int> = flow {
3.      for (i in 1..3) {
4.          Thread.sleep(100)        // 假設我們以消耗 CPU 的方式進行計算
5.          log("Emitting $i")
6.          emit(i)                  // 發射下一個值
7.      }
8.  }.flowOn(Dispatchers.Default)    // 在流建構元中改變消耗 CPU 程式上下文的
                                     // 正確方式
9.  fun main() = runBlocking<Unit> {
10.     val time = measureTimeMillis {
11.         foo()
12.             .buffer()            // 緩衝發射項，無須等待
13.             .collect { value ->
14.                 delay(300)       // 假設我們花費 300 毫秒來處理它
15.                 println(value)
16.             }
17.     }
18.     println("Collected in $time ms")
19. }
```

通道提供了一種在流中傳輸值的方法。Channel 是和 BlockingQueue 十分類
似的概念，不同之處是，它代替了阻塞的 put 操作並提供了暫停的 send，

還替代了阻塞的 take 操作並提供了暫停的 receive。可以透過關閉通道來表明沒有更多的元素將進入通道。在接收者中可以定期使用 for 循環來從通道中接收元素。

```
1.  fun main() = runBlocking<Unit> {
2.      val channel = Channel<Int>()
3.      launch {
4.      // 這裡可能是消耗大量 CPU 運算的非同步邏輯，我們將僅做 5 次整數的平方
            運算並發送
5.          for (x in 1..5) channel.send(x * x)
6.          channel.close() // 結束發送
7.      }
8.      // 這裡列印了 5 次接收到的整數
9.      repeat(5) { println(channel.receive()) }
10.     // 這裡使用 for 循環來列印所有接收到的元素（直到通道關閉）
11.     for (y in channel) println(y)
12.     println("Done!")
13. }
```

除了 Channel() 工廠函數外，還可以用 produce 建置通道。在建置通道時，一個可選的參數是 capacity，它用來指定緩衝區大小。緩衝操作允許發送者在被暫停前發送多個元素，當緩衝區被佔滿的時候將引起阻塞。通道發送和接收操作遵守先進先出原則。

```
1.  fun CoroutineScope.produceSquares(): ReceiveChannel<Int> = produce {
2.      for (x in 1..5) send(x * x)
3.  }
4.  fun main() = runBlocking {
5.      val squares = produceSquares()
6.      // 列印：1 4 9 16 25
7.      squares.consumeEach { println(it) }
8.      println("Done!")
9.  }
```

3.6 小結

本章介紹了 Kotlin 的語法，涵蓋了常用的基礎語法和高階特性。首先介紹了 Kotlin 的基本類型、套件、參考、流程控制、跳躍等；接著介紹了 Kotlin 的類別、屬性、建構函數、介面、泛型；然後介紹了 Kotlin 的函數、Lambda 運算式、集合等，這部分是經常會用到的；最後介紹了程式碼協同，從基礎的實例開始逐步深入介紹程式碼協同，程式碼協同更加輕量級，功能也很強大。這些語法是開發微服務應用的基礎。下一章我們將介紹如何用 Kotlin 整合各種基礎中介軟體。

Kotlin 在常用中介軟體中的應用

本章主要介紹 Kotlin 在常用中介軟體中的應用，透過範例程式，將展示 Kotlin 整 合 Spring Boot、Redis、JPA、QueryDSL、MongoDB、Spring Security、RocketMQ、Elasticsearch、Swagger 的方法。讀者可以掌握使用 Kotlin 操作常用中介軟體的技巧。

4.1 Kotlin 整合 Spring Boot

Spring Boot 是由 Pivotal 團隊開發的，設計的目的是簡化 Spring 應用的初始架設和開發過程。本節介紹 Kotlin 整合 Spring Boot 開發。

4.1.1 Spring Boot 介紹

從 2014 年 4 月發佈 1.0.0.RELEASE 到現在的最新版本 2.2.2.RELEASE，從最初的基於 Spring 4 到現在基於 Spring 5，從同步阻塞程式設計到非同步響應式程式設計，Spring Boot 經歷了數十個 RELEASE 版本，發展迅速，表現穩定，其各版本發行時間如表 4.1 所示。越來越多的企業在生產中使用 Spring Boot 進行企業級應用程式開發。

表 4.1 Spring Boot、Spring 版本的發行時間

時間	Spring Boot 版本	Spring 版本
2014 年	1.0.x	4.0.x.RELEASE
2014—2015 年	1.1.x	4.0.x.RELEASE
2015 年	1.2.x	4.1.x.RELEASE
2015—2016 年	1.3.x	4.2.x.RELEASE
2016—2017 年	1.4.x	4.3.x.RELEASE
2017—2018 年	1.5.x	4.3.x.RELEASE
2018—2019 年	2.0.x	5.0.x.RELEASE
2018—2020 年	2.1.x	5.1.x.RELEASE
2019—2020 年	2.2.x	5.2.x.RELEASE

Spring Boot 基於約定優於設定的思想，讓開發人員不必在設定與邏輯業務之間進行思維的切換。Spring Boot 簡化了 Spring 應用的開發，不再需要 XML 設定檔，使用註釋方式加強了開發效率。Spring Boot 預設設定了很多架構的使用方式，提供 starter 套件，簡化設定，開箱即用。Spring Boot 盡可能地根據專案依賴來自動設定 Spring 架構。Spring Boot 提供了可以直接在生產環境中使用的功能，如效能指標、應用資訊和應用健康檢查。

Spring Boot 內嵌 Tomcat、Jetty、Undertow 等容器，直接用 Jar 套件的方式進行部署，而傳統的 Spring 應用需要用 war 套件方式進行部署。Spring Boot 的部署方法非常簡單，一行指令就可以部署一個 Spring Boot 應用；可以很方便地用 Docker、Kubernetes 進行部署，適用於雲原生應用，使系統的擴充、運行維護更加方便。

Spring Boot 廣泛應用於企業級應用和微服務開發。Spring Cloud 微服務架構就是在 Spring Boot 基礎上開發的。此外，很多開放原始碼專案提供了 Spring Boot 的整合，如 rocketmq- spring-boot-starter，方便使用者使用。

4.1.2 用 Kotlin 開發一個 Spring Boot 專案

在 Spring 網站上建立一個以 Maven 為基礎的 Kotlin Spring Boot 專案。填寫 Group、Artifact，選擇依賴的套件 Spring Web，然後下載到本機，如圖 4.1 所示。

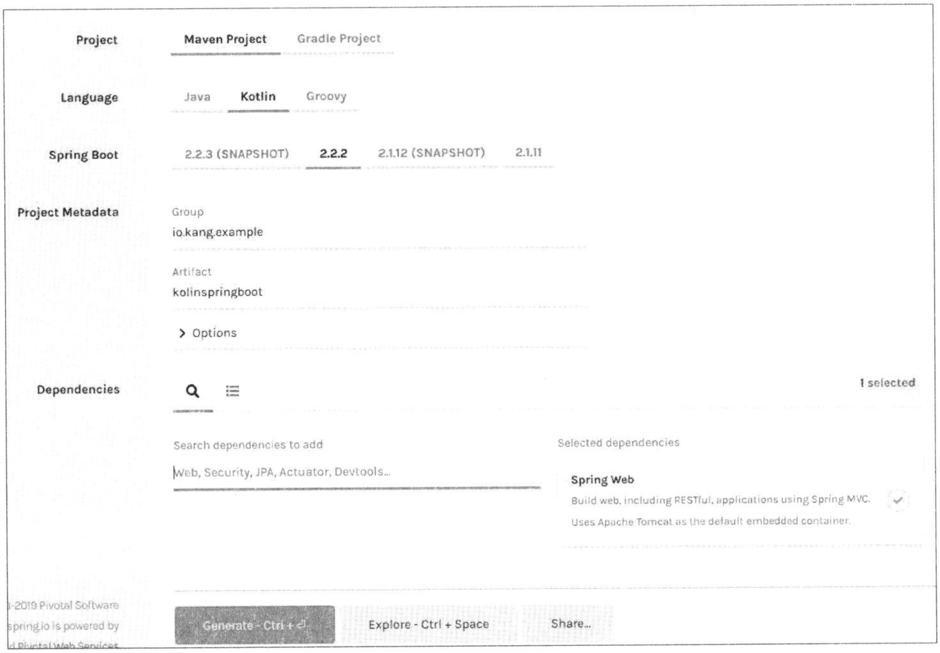

圖 4.1　Spring Initializr

解壓檔案，用 IDEA 開啟這個專案，可以看到 pom 檔案如下：該 pom 檔案定義了父依賴，透過父依賴可以自動找到 dependencies 中相依套件的版本編號；此外，還指定了 Kotlin 的版本是 1.3.61，Spring Boot 的版本是 2.2.2.RELEASE。

```
1.  <?xml version="1.0" encoding="UTF-8"?>
2.  <project xmlns="http://maven.apache.org/POM/4.0.0" xmlns:xsi="http://www.
    w3.org/ 2001/XMLSchema-instance"
3.      xsi:schemaLocation="http://maven.apache.org/POM/4.0.0 https://maven.
```

```
     apache.org/xsd/maven-4.0.0.xsd">
4.       <modelVersion>4.0.0</modelVersion>
5.       <!-- 父 pom，定義套件的依賴 -->
6.       <parent>
7.           <groupId>org.springframework.boot</groupId>
8.           <artifactId>spring-boot-starter-parent</artifactId>
9.           <version>2.2.2.RELEASE</version>
10.          <relativePath/> <!-- lookup parent from repository -->
11.      </parent>
12.      <!-- 子專案相關資訊 -->
13.      <groupId>io.kang.example</groupId>
14.      <artifactId>kolinspringboot</artifactId>
15.      <version>0.0.1-SNAPSHOT</version>
16.      <name>kolinspringboot</name>
17.      <description>Demo project for Spring Boot</description>
18.      <!-- 定義屬性 -->
19.      <properties>
20.          <java.version>1.8</java.version>
21.          <kotlin.version>1.3.61</kotlin.version>
22.      </properties>
23.      <dependencies>
24.          <!-- Spring Boot 啟動套件 -->
25.          <dependency>
26.              <groupId>org.springframework.boot</groupId>
27.              <artifactId>spring-boot-starter</artifactId>
28.          </dependency>
29.          <!-- Kotlin 相關相依套件 -->
30.          <dependency>
31.              <groupId>org.jetbrains.kotlin</groupId>
32.              <artifactId>kotlin-reflect</artifactId>
33.          </dependency>
34.          <dependency>
35.              <groupId>org.jetbrains.kotlin</groupId>
36.              <artifactId>kotlin-stdlib-jdk8</artifactId>
37.          </dependency>
```

```
38.        <dependency>
39.            <groupId>org.springframework.boot</groupId>
40.            <artifactId>spring-boot-starter-test</artifactId>
41.            <scope>test</scope>
42.            <exclusions>
43.                <exclusion>
44.                    <groupId>org.junit.vintage</groupId>
45.                    <artifactId>junit-vintage-engine</artifactId>
46.                </exclusion>
47.            </exclusions>
48.        </dependency>
49.    </dependencies>
50.    <build>
51.        <!— Kotlin 原始程式路徑 -->
52.        <sourceDirectory>${project.basedir}/src/main/kotlin
    </sourceDirectory>
53.        <testSourceDirectory>${project.basedir}/src/test/kotlin
    </testSourceDirectory>
54.        <plugins>
55.            <!— Spring Boot Maven 包裝外掛程式 -->
56.            <plugin>
57.                <groupId>org.springframework.boot</groupId>
58.                <artifactId>spring-boot-maven-plugin</artifactId>
59.            </plugin>
60.            <!— Kotlin Maven 外掛程式 -->
61.            <plugin>
62.                <groupId>org.jetbrains.kotlin</groupId>
63.                <artifactId>kotlin-maven-plugin</artifactId>
64.                <configuration>
65.                    <args>
66.                        <arg>-Xjsr305=strict</arg>
67.                    </args>
68.                    <compilerPlugins>
69.                        <plugin>spring</plugin>
70.                    </compilerPlugins>
```

```
71.                </configuration>
72.                <dependencies>
73.                    <dependency>
74.                        <groupId>org.jetbrains.kotlin</groupId>
75.                        <artifactId>kotlin-maven-allopen</artifactId>
76.                        <version>${kotlin.version}</version>
77.                    </dependency>
78.                </dependencies>
79.            </plugin>
80.        </plugins>
81.    </build>
82.
83. </project>
```

下面用 Kotlin 撰寫一個簡單的 Spring Boot Web 應用:定義一個 Spring
Boot 啟動類別,加上 @SpringBootApplication 註釋;定義一個介面,透過
http://localhost:8080/index 可以造訪這個介面;相關的設定放在 application.
yml 中。

和用 Java 開發 Spring Boot 專案類似,Kotlin 在 main 函數中啟動應用,用
GetMapping 定義一個 get 介面,使用 @RestController 後就不用為每個方
法增加 @ResponseBody 註釋了。Kotlin 的語法更加簡潔。

KotlinSpringbootApplication.kt 的程式如下:

```
1.  @SpringBootApplication
2.  class KotlinSpringbootApplication
3.  // 主函數,啟動類別
4.  fun main(args: Array<String>) {
5.      runApplication<KotlinSpringbootApplication>(*args);
6.  }
```

IndexController.kt 的程式如下:

```
1.  @RestController
```

```
2.  class IndexController {
3.
4.      // 定義 index 介面
5.      @GetMapping("/index")
6.      fun index(): String {
7.          return "Hello, Kotlin for Spring Boot!!"
8.      }
9.  }
```

application.yml 定義應用的設定資訊：

```
1.  server:
2.    port: 8080 # 通訊埠編號
```

透過瀏覽器存取 "index" 介面，顯示 "Hello，Kotlin for Spring Boot!!"。僅透過短短幾行程式就開發了一個簡單的 Kotlin Web 應用，非常便捷。

4.2 Kotlin 整合 Redis

Redis 是用 C 語言撰寫的開放原始碼的、高性能的 key-value 資料庫，讀取速度是 110000 次 / 秒，寫入速度是 81000 次 / 秒，效能極高。本節介紹 Kotlin 整合 Redis 開發和 Redis 相關功能的使用知識。

4.2.1 Redis 介紹

Redis 支援資料持久化，可以將記憶體中的資料儲存到磁碟中，重新啟動的時候可以再次載入使用。Redis 支援豐富的資料類型，包含 string、list、set、zset、hash 等資料結構的儲存。Redis 支援叢集模式，master 節點可以向 slave 節點同步資料。

Redis 為單處理程序單執行緒模式，採用佇列模式將平行處理存取變為串列存取。Redis 的所有操作都是原子性的，支援交易；Redis 還支援 publish/subscribe 通知及 key 過期等特性。

Redis 支援主從複製，當主要資料庫和從資料庫建立主從關係後，向主要資料庫發送 SYNC 指令；主要資料庫收到 SYNC 指令後，開始在後台儲存快照，並將期間收到的指令快取；當快照完成後，主 Redis 將快照檔案和所有快取的寫指令發送給從 Redis；從 Redis 接收後，會載入快照檔案並且執行收到的快取的指令；之後，每當主 Redis 收到寫指令時，就將指令發送給從 Redis，進一步保障資料一致。

Redis 支援高可用，透過一個或多個檢查點 Sentinel 實例組成的 Sentinel 系統可以監控任意個主要伺服器，以及這些主要伺服器的所有從伺服器。當主要伺服器下線時，自動將主要伺服器的某個從伺服器升級為新的主要伺服器，由新的主要伺服器代替已下線的主要伺服器繼續處理指令請求。Sentinel 透過心跳機制監控主從伺服器工作是否正常；當主要伺服器出現問題時，通知相關節點；自動進行故障遷移，自動進行主從切換；支援統一的設定管理，連接者透過 Sentinel 取得主從位址。

Redis 支援的 Java 用戶端有 Jedis 和 Redission 等，官方推薦使用 Redission。Jedis 是 Redis 的 Java 用戶端，提供了對 Redis 指令的全面支援。Redisson 是一個進階的分散式協調 Redis 客服端，實現了分散式和可擴充的 Java 資料結構，如布隆篩檢程式 Bloom filter，集合 Set、ConcurrentMap 等。Redission 和 Jedis 相比，功能較為簡單，不支援字串操作，不支援排序、交易、管線、分區等 Redis 特性。

Redis 的企業級應用有以下場景。

- 分散式快取：快取熱點資料，可以設定過期時間，定時更新快取資料。
- 限時業務：Redis 的 key 可以設定過期時間，到期後 Redis 會刪除這個

key。這一特性可以用於隔一段時間取得手機驗證碼、限時優惠活動資訊等場景。

■ 計數器：Redis 的 incrby 指令可以實現原子性遞增，可以用於高平行處理的秒殺、產生分散式序號、介面限流等。可以限制一個手機號碼發送多少筆簡訊，限制一個介面一分鐘接收多少次請求，一天呼叫多少次等。

■ 排行榜：可使用 Redis 的 sortedSet 進行熱點資料排序。舉例來說，報表平台，可以將統計資料存入 Redis，取得排序好的資料。

■ 分散式鎖：使用 Redis 的 setnx 指令，如果 key 值不存在，成功設定 key 值並快取，同時傳回 1，沒有設定 key 值時傳回 0。利用這個特性，在分散式環境中可以對某個物件加鎖，防止多台實例同時操作。

■ 按讚、好友、附近的人場景：Redis 的 set 不允許存在重復資料，可以判斷某個成員是否在集合內，可以很方便實現尋找共同好友、尋找附近的人等業務場景。

■ 佇列：Redis 的 list 有 push、pop 指令，可以實現佇列操作。

4.2.2 使用 Kotlin 操作 Redis

Spring Boot 整合 Redis 很簡單，在 pom.xml 中增加以下依賴：

```
1.  <!— Spring Boot Redis 相依套件 -->
2.  <dependency>
3.      <groupId>org.springframework.boot</groupId>
4.      <artifactId>spring-boot-starter-data-redis</artifactId>
5.  </dependency>
```

在 application.yml 中增加 Redis 的設定，如 IP 位址、通訊埠、密碼等：

```
1.  spring:
2.    redis:
3.      host: 127.0.0.1      # Redis 伺服器的 IP 位址
4.      port: 6379           # Redis 伺服器的通訊埠編號
5.      password: 123456     # Redis 伺服器密碼
```

Spring Data Redis 提 供 操 作 Redis 的 類 別 RedisTemplate，透 過 植 入 RedisTemplate 可 以 操 作 Redis。RedisTemplate 提 供 opsForValue、opsForList、opsForSet、opsForZset、opsForHash 分 別 處 理 單 一 值、list、set、zset、hash 類型。

定義 RedisData 類別，植入 RedisTemplate<String, String>，key 和 value 都是 String 類型的。saveString 方法可快取單一值：saveStringWithExpire 方法快取單一值，且會在一定時間後過期。getString 方法可以根據 key 取得 value 值。

```kotlin
1.  class RedisData {
2.      // 植入 redisTemplate
3.      @Autowired
4.      private lateinit var redisTemplate: RedisTemplate<String, String>
5.      // 快取到 Redis，key 和 value 都是 String 類型的
6.      fun saveString(key: String, value: String) {
7.          redisTemplate.opsForValue().set(key, value)
8.      }
9.      // 快取到 Redis，expireSecond 秒後過期
10.     fun saveStringWithExpire(key: String, value: String, expireSecond:
    Long){
11.         redisTemplate.opsForValue().set(key, value, Duration.ofSeconds
    (expireSecond))
12.     }
13.     // 從 Redis 取得快設定值
14.     fun getString(key: String): String? {
15.         return redisTemplate.opsForValue().get(key);
16.     }
17. }
```

saveList 方法快取一個 List<String>，將 List 中的元素一個一個發送到快取。saveListWithExpire 會對 key 值設定快取時間。getListValue 方法可以根據 key 值取得 List 的元素，可以指定要取得的元素的範圍。

```
1.  // 將 List 快取到 Redis
2.  fun saveList(key: String, values: List<String>) {
3.      values.forEach { v ->
4.          redisTemplate.opsForList().leftPush(key, v)
5.      }
6.  }
7.  // 將 List 快取到 Redis，expireSecond 秒後過期
8.  fun saveListWithExpire(key: String, values: List<String>, expireSecond:
    Long) {
9.      values.forEach { v ->
10.         redisTemplate.opsForList().leftPush(key, v)
11.     }
12.     redisTemplate.expire(key, expireSecond, TimeUnit.SECONDS)
13. }
14. // 從 Redis 取得 List 中的元素，可以指定元素的範圍
15. fun getListValue(key: String, start: Long, end: Long): List<String>? {
16.     return redisTemplate.opsForList().range(key, start, end);
17. }
```

saveSet 方法可以儲存一組值，並且可以去除重複的值；saveSetWithExpire
方法可以對 key 值設定過期時間；getSetValues 方法可以根據 key 值取得所
有的 value 值；getSetDiff 方法可以根據兩個集合的 key 值取得它們不同的
元素的集合。

```
1.  // 將 set 快取到 Redis，set 中沒有重複元素
2.  fun saveSet(key: String, values: Array<String>) {
3.      redisTemplate.opsForSet().add(key, *values)
4.  }
5.  // 將 set 快取到 Redis，expireSecond 秒後過期
6.  fun saveSetWithExpire(key: String, values: Array<String>, expireSecond:
    Long) {
7.      redisTemplate.opsForSet().add(key, *values)
8.      redisTemplate.expire(key, expireSecond, TimeUnit.SECONDS)
9.  }
10. // 從 Redis 取得快取的 set 集合
```

```
11. fun getSetValues(key: String): Set<String>? {
12.     return redisTemplate.opsForSet().members(key)
13. }
14. // 從 Redis 取得兩個 set 中不相同的元素
15. fun getSetDiff(key1: String, key2: String): Set<String>? {
16.     return redisTemplate.opsForSet().difference(key1, key2)
17. }
```

saveZset 方法可以儲存一組值和值對應的得分，得分可以用於對元素進行
排序。

saveZsetWithExpire 可以對 key 值設定過期時間。getZsetRangeByScore 可
以取得得分在某個區間的值。

```
1.  // 將一組 pair 元素快取到 Redis
2.  fun saveZset(key: String, values: Array<Pair<String, Double>>) {
3.      values.forEach { v ->  redisTemplate.opsForZSet().add(key, v.first,
    v.second)}
4.  }
5.  // 將一組 pair 元素快取到 Redis，expireSecond 秒後過期
6.  fun saveZsetWithExpire(key: String, values: Array<Pair<String, Double>>,
    expireSecond: Long) {
7.      values.forEach { v ->  redisTemplate.opsForZSet().add(key, v.first,
    v.second)}
8.      redisTemplate.expire(key, expireSecond, TimeUnit.SECONDS)
9.  }
10. // 從 Redis 取得在某個區間的值
11. fun getZsetRangeByScore(key: String, minScore: Double, maxScore: Double):
    Set<String>? {
12.     return redisTemplate.opsForZSet().rangeByScore(key, minScore, maxScore)
13. }
```

saveHash 方法可以在 Redis 中儲存一個 map 結構。saveHashWithExpire 方
法可以對 key 設定過期時間。getHashValues 方法可以取得這個 map 中所
有的 value 值。

```kotlin
1.  // 將一個 map 快取到 Redis
2.  fun saveHash(key: String, values: Map<String, String>) {
3.      redisTemplate.opsForHash<String, String>().putAll(key, values);
4.  }
5.  // 將一個 map 快取到 Redis，expireSecond 秒後過期
6.  fun saveHashWithExpire(key: String, values: Map<String, String>,
    expireSecond: Long) {
7.      redisTemplate.opsForHash<String, String>().putAll(key, values);
8.      redisTemplate.expire(key, expireSecond, TimeUnit.SECONDS)
9.  }
10. // 從 Redis 取得 map 中所有的 value 值
11. fun getHashValues(key: String): List<String>? {
12.     return redisTemplate.opsForHash<String, String>().values(key);
13. }
```

可以對上述方法進行測試，定義一個 RedisDataTest 類別測試上述方法，植入 RedisData。透過測試可以看到，在 Redis 中快取單一 key，如果設定過期時間，超過該時間後取得的 value 是 null。

```kotlin
1.  @SpringBootTest
2.  @RunWith(SpringRunner::class)
3.  class RedisDataTest {
4.      @Autowired
5.      lateinit var redisData: RedisData
6.      // 測試將一個字串快取到 Redis，2 秒後自動過期
7.      @Test
8.      fun `save redis key value with expire time`() {
9.          runBlocking {
10.             redisData.saveStringWithExpire("hello1", "helloWorld1", 2L)
11.             kotlinx.coroutines.delay(2 * 1000L)
12.         }
13.         println("get key")
14.         val value = redisData.getString("hello1")
15.         Assert.assertNull(value)
16.     }
```

```
17.        // 測試將一個值快取到 Redis，並從快取中取出這個值
18.        @Test
19.        fun `save and get redis value`() {
20.            val key = "hello"
21.            redisData.saveString(key, "helloWorld")
22.            val value = redisData.getString(key)
23.            Assert.assertEquals(value, "helloWorld")
24.        }
25.    }
```

在 Redis 中快取一個 List，可以取得某個範圍的元素，當指定 range(0, 1)
時可以取得兩個元素。

```
1.    // 測試將一個 List 快取到 Redis，2 秒後自動過期
2.    @Test
3.    fun `save redis list with expire time`() {
4.        runBlocking {
5.            val values = arrayListOf("hi1", "hi2", "hi3")
6.            redisData.saveListWithExpire("listKey1", values, 2L)
7.            kotlinx.coroutines.delay(2 * 1000L)
8.        }
9.        val values = redisData.getListValue("listKey1", 0L, 3L)
10.       Assert.assertEquals(values?.size, 0)
11. }
12. // 測試將一個 List 快取到 Redis，並從 Redis 取得這個 List 的元素
13. @Test
14. fun `save and get list values`() {
15.       val key = "listKey"
16.       redisData.saveList(key, arrayListOf("hi1", "hi2", "hi3"))
17.       val values = redisData.getListValue("listKey", 0L, 1L)
18.       Assert.assertEquals(2, values?.size)
19. }
```

在 Redis 中儲存兩組元素，其中兩組元素都有重複的，從下面的程式中可
以看到這兩組元素中不同的元素只有 1 個。

```
1.  // 測試將一個 Set 快取到 Redis，2s 後自動過期
2.  @Test
3.  fun `save redis set with expire time`() {
4.      runBlocking {
5.          val values = arrayOf("hello", "hello", "world")
6.          redisData.saveSetWithExpire("setKey3", values, 2L)
7.          kotlinx.coroutines.delay(2 * 1000L)
8.      }
9.      val values = redisData.getSetValues("setKey3");
10.     Assert.assertEquals(values?.size, 0)
11. }
12. // 測試將一個 Set 快取到 Redis，從 Redis 取出這兩個 Set 中不同的元素
13. @Test
14. fun `save and get redis two set diff`() {
15.     redisData.saveSet("setKey1", arrayOf("hello", "hello", "world", "wide"))
16.     redisData.saveSet("setKey2", arrayOf("hello", "hello", "world", "women"))
17.     val diffSet = redisData.getSetDiff("setKey1", "setKey2")
18.     Assert.assertEquals(diffSet?.size, 1)
19. }
```

在 Redis 中儲存一個 zset，可以指定取得某個分數範圍的元素，程式如下所示，可以取得兩個元素。

```
1.  // 測試將一組 pair 快取到 Redis，2 秒後自動過期
2.  @Test
3.  fun `save redis zset with expire time`() {
4.      runBlocking {
5.          val values = arrayOf(Pair("xiaoming",98.0), Pair("xiaoli", 90.0),
    Pair("wangming", 100.0))
6.          redisData.saveZsetWithExpire("zsetKey2", values, 2L)
7.          kotlinx.coroutines.delay(2_000)
8.      }
9.      val values = redisData.getZsetRangeByScore("zsetKey2", 95.0, 99.0);
10.     Assert.assertEquals(values?.size, 0)
11. }
```

```
12. // 測試將一組 pair 快取到 Redis，從 Redis 取得分數範圍在 95.0 到 100.0 的元素
13. @Test
14. fun `saven and get redis zset values by score`() {
15.     redisData.saveZset("zsetKey1", arrayOf(Pair("xiaoming",98.0),
    Pair("xiaoli", 90.0), Pair("wangming", 100.0)))
16.     val values = redisData.getZsetRangeByScore("zsetKey1", 95.0, 100.0)
17.     Assert.assertEquals(2, values?.size)
18. }
```

在 Redis 中儲存一個 hash，測試以下程式，可以取得 map 的 value 的數量是 2。

```
1.  // 測試將一個 map 快取到 Redis，2 秒後自動過期
2.  @Test
3.  fun `save redis hashs with expire time`() {
4.      runBlocking {
5.          val aMap = mapOf(Pair("key1","value1"), Pair("key2", "value2"))
6.          redisData.saveHashWithExpire("hashKey2", aMap, 2L)
7.          kotlinx.coroutines.delay(2_000)
8.      }
9.      val values = redisData.getHashValues("hashKey2")
10.     Assert.assertEquals(0, values?.size)
11. }
12. // 測試將一個 map 快取到 Redis，取得 map 所有的 value 值
13. @Test
14. fun `save and get redis hash values`() {
15.     val aMap = mapOf(Pair("key1","value1"), Pair("key2", "value2"))
16.     redisData.saveHash("hashKey1", aMap)
17.     val values = redisData.getHashValues("hashKey1")
18.     Assert.assertEquals(2, values?.size)
19. }
```

4.3 Kotlin 整合 JPA、QueryDSL

JPA 全稱是 Java Persistence API，可以透過註釋描述物件和資料庫表之間的關係，將實體物件持久化到資料庫中。QueryDSL 彌補了 JPA 多表動態查詢的不足，並且和 JPA 高度整合。本節介紹 Kotlin 整合 JPA、QueryDSL 的開發。

4.3.1 JPA、QueryDSL 介紹

JPA 提供了 ORM（物件關係對映）對映中繼資料，用中繼資料描述物件和表之間的關係，支援 @Entity、@Table、@Column、@Transient 等註釋。可以將開發者從煩瑣的 JDBC 和 SQL 中解脫出來，使用 API 執行 CRUD 操作，架構會隱藏處理細節。JPA 是一種標準，定義了一些介面，Hibernate 是實現了 JPA 介面的 ORM 架構。Spring Data JPA 是 Spring 提供的簡化 JPA 開發的架構，只需要寫 DAO 層介面，就可以在不實現介面的情況下，實現對資料庫的操作。其同時還支援分頁、排序、複雜查詢及原生 SQL 等。Spring Data JPA 是對 JPA 標準的再次封裝，底層使用的還是 Hibernate 的 JPA 技術實現。MyBatis 需要在 XML 檔案中定義 SQL，JPA 更加簡潔。

JPA 具有以下特點。

- 標準化：JPA 為不同資料庫提供了相同的 API，這確保了以 JPA 開發為基礎的企業應用經過少量的修改就能夠在不同的 JPA 架構下執行。

- 簡單好用，整合方便：JPA 的主要目標之一是提供更加簡單的程式設計模型，在 JPA 架構下建立實體和建立 Java 類別一樣簡單，只需使用 javax.persistence.Entity 進行註釋；JPA 的架構和介面也都非常簡單。

- 可媲美 JDBC 的查詢能力：JPA 的查詢語言是物件導向的，JPA 定義了獨特的 JPQL，支援批次更新和修改、JOIN、GROUP BY、HAVING 等通常只有 SQL 才能夠提供的進階查詢特性，甚至還能支援子查詢。

- 支援物件導向的進階特性：JPA 中支援物件導向的進階特性，如類別之間的繼承、多形和類別之間的複雜關係，大幅地使用物件導向的模型。

對於單表查詢，JPA 使用起來非常方便。但當有關多表動態查詢時，JPA 沒有那麼靈活，常常需要使用 @Query 註釋，如果在這個註釋中寫 SQL，那麼 SQL 的可讀性比較差。QueryDSL 是以各種 ORM 上為基礎的通用架構，使用 QueryDSL 的 API 類別庫，可建置類型安全的 SQL 查詢。以 QueryDSL 為基礎可以在任何支援的 ORM 架構或 SQL 平台上以一種通用的 API 方式來建置查詢。目前，QueryDSL 支援的平台包含 JPA、JDO、SQL、Java Collections、RDF、Lucene、Hibernate Search。QueryDSL 的語法和 SQL 十分類似，程式可讀性強，簡潔優美。QueryDSL 使用 API 建置查詢，可以安全地參考欄位型態和屬性。

QueryDSL 支援程式自動完成，由於它基於 Java API，因此可以利用 Java IDE 的程式自動補全功能；QueryDSL 幾乎可以避免所有的 SQL 語法錯誤；QueryDSL 採用 Domain 類型的物件和屬性來建置查詢，因此是類型安全的，不會因為條件類型而出現問題；QueryDSL 採用純 Java API 建置 SQL，便於程式重構；QueryDSL 可以更輕鬆地定義增量查詢。

QueryDSL 並不使用現有 POJO 建置查詢，而是根據現有的設定產生對應的 Domain Model 建置查詢。QueryDSL 外掛程式會基於 JPA 的 POJO 實體自動產生查詢實體，命名方式是 Q+ 對應實體名稱。

下面將透過實際的實例介紹如何用 Kotlin 操作 JPA 和 QueryDSL。

4.3.2 使用 Kotlin 操作 JPA、QueryDSL

Kotlin 整合 JPA、QueryDSL 需要在 pom.xml 中增加以下依賴：

```
1.  <!— Spring Boot JPA 相依套件 -->
2.  <dependency>
3.      <groupId>org.springframework.boot</groupId>
4.      <artifactId>spring-boot-starter-data-jpa</artifactId>
5.  </dependency>
6.  <!— QueryDSL 相依套件 -->
7.  <dependency>
8.      <groupId>com.querydsl</groupId>
9.      <artifactId>querydsl-jpa</artifactId>
10. </dependency>
11. <dependency>
12.     <groupId>com.querydsl</groupId>
13.     <artifactId>querydsl-apt</artifactId>
14. </dependency>
15. <!— MySQL 驅動相依套件 -->
16. <dependency>
17.     <groupId>mysql</groupId>
18.     <artifactId>mysql-connector-java</artifactId>
19. </dependency>
```

此外，還需要引用 QueryDSL Maven 外掛程式，這個外掛程式可以根據 entity 產生 QueryDSL 的 entity。

```
1.  <!— QueryDSL Maven 外掛程式 -->
2.  <plugin>
3.      <groupId>com.querydsl</groupId>
4.      <artifactId>querydsl-maven-plugin</artifactId>
5.      <executions>
6.          <execution>
7.              <phase>compile</phase>
8.              <goals>
9.                  <goal>jpa-export</goal>
```

```
10.            </goals>
11.            <configuration>
12.               <targetFolder>target/generatedsources/kotlin</targetFolder>
                  <packages>io.kang.example.entity</packages>
13.            </configuration>
14.         </execution>
15.      </executions>
16. </plugin>
```

在 application.yml 中增加以下設定，包含設定資料庫的位址、使用者名
稱、密碼等，jpa.hibernate.ddl-auto 為 update 可以在程式啟動時根據 entity
自動在資料庫中建表。

```
1.  spring:
2.    datasource:
3.      password: 123456# 資料庫連接密碼
4.      username: root          # 資料庫連接使用者名稱
5.      url: jdbc:mysql://127.0.0.1:3306/video?characterEncoding=
    utf-8&serverTimezone=UTC   # 資料庫連接 URL
6.    jpa:
7.      hibernate:
8.        ddl-auto: update      # 如果服務啟動時表格式不一致則更新表
```

定義一個 entity，名為 User。然後定義一個使用 JPA 操作 User 的介面
UserRepository，這個介面繼承自 CrudRepository 介面，可根據定義的方法
自動產生底層 SQL。

User.kt 的程式如下所示：

```
1.  // 定義 user 實體
2.  @Entity
3.  data class User(
4.      // 主鍵，自動增加
5.       @Id
6.      @GeneratedValue(strategy = GenerationType.AUTO)
```

```kotlin
7.          val id: Long,
8.          val userName: String,
9.          val password: String,
10.         val email: String,
11.         val age: Int,
12.         val height: Double,
13.         val address: String,
14.         val education: EducationLevel,
15.         val income: Double
16. )
17. // 列舉類別
18. enum class EducationLevel {
19.     XIAOXUE, GAOZHONG, BENKE, YANJIUSHENG, BOSHI
20. }
```

UserRepository.kt 定義了對 User 表的操作方法。CrudRepository 定義了保存單筆 / 全部、根據主鍵查詢、某筆記錄是否存在、查詢全部、計數、根據主鍵刪除、刪除單筆記錄、批次刪除、刪除全部這些方法。可以自訂方法擴充操作。

```kotlin
1.  interface UserRepository: CrudRepository<User, Long> {
2.      // 根據 userName 和 password 尋找使用者
3.      fun findByUserNameAndPassword(userName: String, password: String): User?
        // 根據 userName 進行模糊查詢
4.      fun findByUserNameLike(userName: String): List<User>?
5.      // 尋找收入大於 income 的使用者
6.      fun findByIncomeGreaterThan(income: Double): List<User>?
7.      // 尋找使用者名稱包含 userName 的使用者
8.      fun findByUserNameContains(userName: String): List<User>?
9.      // 根據 userName 和 email 刪除使用者
10.     @Transactional(rollbackFor = [Exception::class])
11.     fun deleteByUserNameAndEmail(userName: String, email: String): Int?
12.     // 增加使用者
13.     fun save(use: User)
14. }
```

UserRepository 提供了一些範例方法,如根據條件尋找、模糊查詢、查詢大於某個值的記錄、查詢包含某個值的元素、條件刪除、儲存記錄等。此外,可以自訂方法,在方法名稱上指定 SQL 運算符號(插入、查詢、刪除等)、屬性、屬性的運算子等。

UserRepositoryTest.kt 對以上方法進行單元測試。testSaveUsers 方法儲存三筆 User 記錄。

```kotlin
1.   @SpringBootTest
2.   @TestMethodOrder(MethodOrderer.OrderAnnotation::class)
3.   @ExtendWith(SpringExtension::class)
4.   class UserRepositoryTest {
5.       @Autowired
6.       lateinit var userReposiroty: UserRepository
7.       // 初始化三筆測試記錄
8.       @Test
9.       @Order(1)
10.      fun testSaveUsers() {
11.          userReposiroty.deleteAll()
12.          var users = arrayOf(
13.                  User(0, "test01", "test01", "test01@qq.com", 45, 175.5,
     "Shanghai", EducationLevel.BOSHI, 50000.0),
14.                  User(1, "test02", "test02", "test02@qq.com", 36, 170.5,
     "Shanghai", EducationLevel.YANJIUSHENG, 20000.0),
15.                  User(2, "test03", "test03", "test03@qq.com", 26, 165.5,
     "Shanghai", EducationLevel.YANJIUSHENG, 10000.0)
16.          )
17.          userReposiroty.saveAll(users.asList().asIterable())
18.          val users1 = userReposiroty.findAll()
19.          assertEquals(3, users1.toList().size)
20.      }
```

testFindByUserNameAndPassword 方法根據使用者名稱、密碼 ——"test01" 和 "test01"——尋找到一筆 User 記錄。

```
1.  // 測試尋找使用者名稱是 "test01"、密碼是 "test01" 的使用者
2.  @Test
3.  @Order(2)
4.  fun testFindByUserNameAndPassword() {
5.      val user = userReposiroty.findByUserNameAndPassword("test01", "test01")
6.      assertEquals("test01@qq.com", user?.email)
7.      assertEquals(45, user?.age)
8.      assertEquals(175.5, user?.height)
9.      assertEquals(EducationLevel.BOSHI, user?.education)
10. }
```

testFindByUserNameLike 方法根據使用者名稱進行模糊查詢，查詢使用者名稱中包含 "test" 字串的記錄。

```
1.  // 測試尋找使用者名稱包含 "test" 字串的使用者
2.  @Test
3.  @Order(3)
4.  fun testFindByUserNameLike() {
5.      val users = userReposiroty.findByUserNameLike("%test%")
6.      assertEquals(3, users?.size)
7.  }
```

testFindByIncomeGreaterThan 方法查詢收入大於 10000.0 的記錄，此外，還支援大於或等於、等於、小於、小於或等於這些運算符號。

```
1.  // 測試尋找收入大於 10 000.0 的使用者
2.  @Test
3.  @Order(4)
4.  fun testFindByIncomeGreaterThan() {
5.      val users = userReposiroty.findByIncomeGreaterThan(10000.0)
6.      assertEquals(2, users?.size)
7.  }
```

testFindByUserNameContains 方法尋找使用者名稱包含 "test" 字串的記錄。

```
1.  // 測試尋找使用者名稱包含 "test" 字串的使用者
2.  @Test
3.  @Order(5)
4.  fun testFindByUserNameContains() {
5.      val users = userReposiroty.findByUserNameContains("test")
6.      assertEquals(3, users?.size)
7.  }
```

testDeleteByUserNameAndEmail 方法根據使用者名稱和電子郵件刪除記錄，刪除使用者名稱是 "test01"、電子郵件是 "test01@qq.com" 的記錄。

```
1.  // 測試根目錄據使用者名稱和電子郵件刪除使用者
2.  @Test
3.  @Order(6)
4.  fun testDeleteByUserNameAndEmail() {
5.      userReposiroty.deleteByUserNameAndEmail("test01", "test01@qq.com")
6.      val users = userReposiroty.findAll().toList()
7.      assertEquals(2, users.size)
8.  }
```

testSave 方法保存單筆記錄。

```
1.  // 測試儲存一筆使用者記錄
2.  @Test
3.  @Order(7)
4.  fun testSave() {
5.      val user = User(3, "test04", "test04", "test04@qq.com", 26, 165.5,
    "Shanghai", EducationLevel.YANJIUSHENG, 10000.0)
6.      userReposiroty.save(user)
7.      val users = userReposiroty.findAll().toList()
8.      assertEquals(3, users.size)
9.  }
```

植入 JPAQueryFactory，可以使用 QueryDSL 操作資料庫。用變數 predicate 儲存 where 條件，queryFactory 提供查詢、刪除和更新的方法。

JPAQueryFactory 操作的是 QUser 物件，是使用外掛程式在 User 基礎上產生的。

testFindByUserNameAndPassword 方法展示了使用 queryDSL 查詢使用者名稱和密碼分別是 "test01" 和 "test01" 的記錄。

```kotlin
1.  @SpringBootTest
2.  @TestMethodOrder(MethodOrderer.OrderAnnotation::class)
3.  @ExtendWith(SpringExtension::class)
4.  class UserDomainRepositoryTest {
5.      // 植入 queryFactory
6.      @Autowired
7.      lateinit var queryFactory: JPAQueryFactory
8.      // 測試根目錄據使用者名稱、密碼尋找使用者
9.      @Test
10.     @Order(1)
11.     fun testFindByUserNameAndPassword() {
12.         val qUser = QUser.user
13.         val predicate = qUser.userName.eq("test01").and(qUser.password.
    eq("test01"))
14.         val user = queryFactory.selectFrom(qUser).where(predicate).
    fetchOne()
15.         Assert.assertEquals("test01@qq.com", user?.email)
16.         Assert.assertEquals(45, user?.age)
17.         Assert.assertEquals(175.5, user?.height)
18.         Assert.assertEquals(EducationLevel.BOSHI, user?.education)
19.     }
20. }
```

testFindByUserNameLike 方法展示了使用 queryDSL 對使用者名稱進行模糊查詢。

```kotlin
1.  // 測試使用者名稱包含 "test" 字串的使用者
2.  @Test
3.  @Order(2)
4.  fun testFindByUserNameLike() {
```

```
5.        val qUser = QUser.user
6.        val predicate = qUser.userName.like("%test%")
7.        val users = queryFactory.selectFrom(qUser).where(predicate).fetch()
8.        Assert.assertEquals(3, users?.size)
9.    }
```

testFindByIncomeGreaterThan 方法展示了使用 queryDSL 查詢收入大於 10 000.0 的記錄。

```
1.    // 測試收入大於 10 000.0 的使用者
2.    @Test
3.    @Order(3)
4.    fun testFindByIncomeGreaterThan() {
5.        val qUser = QUser.user
6.        val predicate = qUser.income.gt(10000.0)
7.        val users = queryFactory.selectFrom(qUser).where(predicate).fetch()
8.        Assert.assertEquals(2, users.size)
9.    }
```

testFindByUserNameContains 方法展示了使用 queryDSL 查詢使用者名稱包含 "test" 的記錄。

```
1.    // 測試使用者名稱包含 "test" 的使用者
2.    @Test
3.    @Order(4)
4.    fun testFindByUserNameContains() {
5.        val qUser = QUser.user
6.        val predicate = qUser.userName.contains("test")
7.        val users = queryFactory.selectFrom(qUser).where(predicate).fetch()
8.        Assert.assertEquals(3, users.size)
9.    }
```

testDeleteByUserNameAndEmail 方法展示了使用 queryDSL 刪除使用者名稱和電子郵件分別是 "test01" 和 "test01@qq.com" 的記錄。由於有關刪除操作，因此需要交易註釋。

```
1.   // 測試根據使用者名稱和電子郵件刪除使用者
2.   @Test
3.   @Order(5)
4.   @Transactional
5.   @Rollback(false)
6.   fun testDeleteByUserNameAndEmail() {
7.       val qUser = QUser.user
8.       val predicate = qUser.userName.eq("test01").and(qUser.email.eq
     ("test01@qq.com"))
9.       queryFactory.delete(qUser).where(predicate).execute()
10.      val users = queryFactory.selectFrom(qUser).fetch()
11.      Assert.assertEquals(2, users.size)
12. }
```

testUpdateEmailByUserName 方法展示了使用 queryDSL 把使用者名稱是
"test02" 的記錄的電子郵件更新為 "test02@yy.com"。

```
1.   // 測試更新使用者名稱是 test02 的使用者的電子郵件
2.   @Test
3.   @Order(6)
4.   @Transactional
5.   @Rollback(false)
6.   fun testUpdateEmailByUserName() {
7.       val qUser = QUser.user
8.       val predicate = qUser.userName.eq("test02")
9.       queryFactory.update(qUser).set(qUser.email, "test02@yy.com").where
     (predicate).execute()
10.      val user = queryFactory.selectFrom(qUser).where(predicate).fetchOne()
11.      Assert.assertEquals("test02@yy.com", user?.email)
12. }
```

4.4 Kotlin 整合 MongoDB

MongoDB 是用 C++ 撰寫的高性能非關聯式資料庫,是以分散式檔案儲存為基礎的開放原始碼資料庫系統。MongoDB 為企業級應用提供了可擴充的資料儲存解決方案。本節介紹 MongoDB 的背景知識以及如何使用 Kotlin 整合 MongoDB 開發。

4.4.1 MongoDB 介紹

MongoDB 將資料儲存為一個文件,資料結構由鍵值對組成,類似 JSON 物件。欄位值可以包含 null、布林值、整數、浮點數、字串、唯一 id、日期、正規表示法、程式、undefined、陣列及其他文件。

MongoDB 是一個針對文件的資料庫,操作起來比較簡單。它可以對任何屬性建立索引,以加強檢索速度。它還支援豐富的查詢運算式,查詢指令使用 JSON 形式的標記,可輕易查詢文件中內嵌的物件以及陣列。它使用 update 指令更新文件的資料。它支援 map/reduce 操作,可對資料進行批次處理和聚合操作。它使用 map 函數呼叫 emit(key, value) 檢查集合中所有的記錄,將 key 和 value 傳給 reduce 函數進行處理。mongoDB 允許在服務端執行指令稿,用 JavaScript 撰寫函數,直接在服務端執行,也可以把函數的定義儲存在服務端,用戶端直接呼叫該函數。MongoDB 支援 Ruby、Python、Java、Kotlin、C++、PHP、C# 等程式語言。

一台 MongoDB 的儲存容量有限,當需要擴充儲存容量時,可以進行水平擴充。MongoDB 透過 Sharded cluster 保障可擴充性,Sharded 是指複製集或單一 Mongos 節點。Sharded cluster 由 Shard、Mongos 和 Config server 等 3 個元件組成。Mongos 是 Sharded cluster 的存取入口。Mongos 本身並不持久化資料,Sharded cluster 所有的中繼資料都會儲存到 Config server,

而使用者的資料則會分散儲存到各個 Shard。Mongos 啟動後，會從 Config server 載入中繼資料，開始提供服務，將使用者的請求正確路由到對應的 Shard。Sharded cluster 支援將單一集合的資料分散儲存在多個 Shard 上，使用者可以指定根據集合內文件的某個欄位即 shard key 來分佈資料，目前主要支援兩種資料分佈策略 —— 範圍分片（Range based sharding）或 hash 分片（Hash based sharding）。Mongos 會根據請求類型及 shard key 將請求路由到對應的 Shard。Config server 儲存 Sharded cluster 的所有中繼資料，所有的中繼資料都儲存在 Config 資料庫中。config.shards 集合儲存各個 Shard 的資訊，可透過 addShard、removeShard 指令動態地從 Sharded cluster 裡增加或移除 Shard。config.databases 集合儲存所有資料庫的資訊，包含資料庫是否開啟分片和 primary shard 的資訊等，對於資料庫內沒有開啟分片的集合，所有的資料都會儲存在資料庫的 primary shard 上。

MongoDB 4.0 以上版本支援交易，MongoDB 4.0 支援複本集交易，存在單一文件最大 16MB、交易執行時間不能過長的限制。MongoDB 4.2 版本支援分散式交易，支援修改分片 key 的內容。交易的執行過程是：取得階段，開啟交易，獲得集合，多個集合操作，回覆交易，提交交易。

MongoDB 適用於儲存記錄檔，不需要像關聯式資料庫那樣設計統一的表結構，儲存更加靈活方便。儲存監控資料、增加欄位都不需要修改表結構，使用成本低。一些 O2O 快遞應用，將司機、商家、位置資訊儲存在 MongoDB 中，透過地理位置查詢附近的商家、司機等。在證券交易類別應用中，將使用者的交易資料儲存到 MongoDB 中，根據這些交易資料可定時產生市場行情。

4.4.2 使用 Kotlin 操作 MongoDB

Kotlin 整合 MongoDB 需要在 pom.xml 中增加以下依賴:

```
1.  <!-- Spring Boot MongoDB 相依套件 -->
2.  <dependency>
3.      <groupId>org.springframework.boot</groupId>
4.      <artifactId>spring-boot-starter-data-mongodb</artifactId>
5.  </dependency>
```

在 application.yml 中增加以下設定,設定 MongoDB 的 URL,我們使用的是單機 MongoDB:

```
1.  spring:
2.    data:
3.      mongodb:
4.        uri: mongodb://exchange:123456@127.0.0.1:27017/exchange
    #mongodb 連接 URL
```

Person.kt 定義了集合和實體的對映關係,Student 繼承了 Person,Student 集合具有 personId、name、address、age、date、likeSport、likeBook、school 這些屬性。

```
1.  // 定義 Person 實體類別
2.  open class Person(
3.      @Id
4.      var personId: Long = 0L,
5.      var name: String = "",
6.      var address: String = "",
7.      var age: Int = 0,
8.      var date: Date = Date()
9.  )
10. // 定義一個 Student 實體類別,對映為 MongoDB 中的 Student 集合
11. @Document(collection="student")
12. data class Student(
13.     val likeSport: String,
```

```
14.        val likeBook: String,
15.        val school: String
16. ): Person()
```

StudentRepository.kt 定義了操作 Student 集合的介面,它繼承了 MongoRepository。MongoRepository 提供了批次 / 單筆儲存、全部查詢、條件查詢等基本方法。它還可以擴充集合的操作方法,方法名稱由操作方法、屬性和運算符號組成;也可以使用 @DeleteQuery、@Query 註釋自訂原生的刪除、查詢敘述。

```
1.  interface StudentRepository: MongoRepository<Student, Long> {
2.      // 根據 personId 尋找 student
3.      fun findByPersonId(personId: Long): Student?
4.      // 分頁尋找年齡大於或等於 age,模糊比對 name 的 student
5.      fun findByNameRegexAndAgeGreaterThanEqual(name: String, age: Int,
    pageable: Pageable): Page<Student>
6.      // 使用原生 SQL 刪除年齡在 age1 到 age2 範圍的資料
7.      @DeleteQuery(value = "{\"age\":{\"\$gte\":?0,\"\$lte\":?1}}")
8.      fun deleteByAgeIn(age1: Int, age2: Int)
9.      // 使用原生 SQL 分頁尋找年齡大於或等於 age,模糊比對 name 的 student
10.     @Query(value = "{\"name\":{\"\$regex\":?0},\"age\":{\"\$gte\":?1}}")
11.     fun findByAgeIndividual(name: String ,age: Int, pageable: Pageable):
    Page<Student>
12. }
```

StudentRepositoryTest.kt 對 StudentRepository 定義的方法進行了測試。此外,也可以使用 MongoTemplate 來操作集合。saveStudents 方法向 Student 集合中儲存了 16 筆記錄。

```
1.  @SpringBootTest
2.  @TestMethodOrder(MethodOrderer.OrderAnnotation::class)
3.  @ExtendWith(SpringExtension::class)
4.  class StudentRepositoryTest {
5.      @Autowired
6.      lateinit var studentRepository: StudentRepository
```

```
7.      @Autowired
8.      lateinit var mongoTemplate: MongoTemplate
9.      // 測試儲存 student 資料，初始化 16 筆測試記錄
10.     @Test
11.     @Order(1)
12.     fun saveStudents() {
13.         val student = Student(" 籃球 ","Kotlin 程式設計 "," 上海中學 ")
14.         student.personId = 20180101L
15.         student.age = 22
16.         student.name = " 張三 "
17.
18.         studentRepository.deleteAll()
19.
20.         for(i in 0..15) {
21.             student.age = 22 - i
22.             student.personId = 20180101L + i
23.             studentRepository.save(student)
24.         }
25.     }
26. }
```

testFindByNameAndAge 方法對名字是「張三」、年齡大於或等於 20 的記錄進行分頁查詢，每頁 5 筆記錄，查詢第 0 頁。

```
1.  // 測試尋找名字是 " 張三 "、年齡大於或等於 20 的 student，查到 3 筆
2.  @Test
3.  @Order(2)
4.  fun testFindByNameAndAge() {
5.      val pageable = PageRequest.of(0, 5)
6.
7.      val studentPage = studentRepository.findByNameRegexAndAgeGreaterThanEqual
    (" 張三 ", 20, pageable)
8.
9.      Assert.assertEquals(3, studentPage.content.size)
10. }
```

testFindByNameAndAge1 方法使用原生的查詢敘述對姓名是「張三」、年齡大於或等於 10 的記錄進行分頁查詢。用 JSON 格式定義查詢準則：{"name":{"$regex":" 張三 "},"age":{"$gte":10}}，而 testFindByNameAndAge 是 Mongo JPA 根據定義的方法自動產生查詢敘述。

```
1.  // 測試使用原生 SQL 尋找名字是 " 張三 "、年齡大於或等於 10 的 student，查到 5 筆
2.  @Test
3.  @Order(3)
4.  fun testFindByNameAndAge1() {
5.      val pageable = PageRequest.of(0, 5)
6.      val studentPage = studentRepository.findByAgeIndividual(" 張三 ", 10,
    pageable)
7.      Assert.assertEquals(5, studentPage.content.size)
8.  }
```

testDeleteByAgeIn 方法刪除年齡大於或等於 20 且小於或等於 22 的記錄。

```
1.  // 測試刪除年齡大於或等於 20 且小於或等於 22 的 student
2.  @Test
3.  @Order(4)
4.  fun testDeleteByAgeIn() {
5.      studentRepository.deleteByAgeIn(20, 22)
6.      val students = studentRepository.findAll()
7.      Assert.assertEquals(13, students.size)
8.  }
```

testSortByAge 方法使用 mongoTemplate 操作集合，透過 aggregation 定義對集合的實際操作：指定操作的集合，根據年齡降冪尋找，只取 1 筆記錄。

```
1.  // 測試使用 mongoTemplate 根據 age 降冪尋找，只取 1 筆記錄
2.  @Test
3.  @Order(5)
4.  fun testSortByAge() {
5.      val aggregation = Aggregation.newAggregation(Student::class.java,
```

```
      sort(Sort.Direction.DESC, "age"), limit(1))
6.    val student = mongoTemplate.aggregate(aggregation, Student::class.
java).mappedResults
7.    Assert.assertEquals(19, student[0].age)
8. }
```

testFindCount 方法查詢集合 Student 的總數。

```
1. // 測試使用 mongoTemplate 尋找 Student 集合的總數
2. @Test
3. @Order(6)
4. fun testFindCount() {
5.    val n = mongoTemplate.db.getCollection("student").countDocuments()
6.    Assert.assertEquals(13, n)
7. }
```

testFindByName 方法先取得集合 Student，傳回滿足姓名等於「張三」、年齡大於或等於 19 的記錄的總數。

```
1. // 測試使用 mongoTemplate 尋找名字是 " 張三 "、年齡大於或等於 19 的 student
   的總數
2. @Test
3. @Order(7)
4. fun testFindByName() {
5.    val n = mongoTemplate.db.getCollection("student").find()
6.           .filter(eq("name", " 張三 "))
7.           .filter(gte("age", 19)).count()
8.    Assert.assertEquals(1, n)
9. }
```

testFindByNameLimit 方法取得集合 Student，過濾姓名等於「張三」的記錄，取前 5 筆，並傳回總數。

```
1. // 測試使用 mongoTemplate 尋找名字是 " 張三 " 的記錄，取 5 筆記錄
2. @Test
3. @Order(8)
```

```
4.  fun testFindByNameLimit() {
5.      val n = mongoTemplate.db.getCollection("student").find()
6.              .filter(eq("name", " 張三 "))
7.              .limit(5)
8.              .count()
9.      Assert.assertEquals(5, n)
10. }
```

testFindByNameCurse 方法對集合 Student 操作，過濾姓名等於「張三」的
記錄，開啟一個游標，透過游標可以檢查這些記錄。

```
1.  // 測試使用 mongoTemplate 尋找 name 是 " 張三 " 的記錄，並用游標檢查元素
2.  @Test
3.  @Order(9)
4.  fun testFindByNameCurse() {
5.      val studentCursor = mongoTemplate.db.getCollection("student").find()
6.              .filter(eq("name", " 張三 "))
7.              .cursor()
8.      while(studentCursor.hasNext()){
9.          Assert.assertEquals(" 張三 ", studentCursor.next()["name"])
10.     }
11.     studentCursor.close()
12. }
```

testFindObjectId 方 法 重 點 測 試 MongoDB 的 主 鍵 _id，_id 的 類 型 是
ObjectId。ObjectId 是一個分散式 id，使用 12 位元組的儲存空間，是一
個 24 位元的字串。前 4 位元組是一個時間戳記，單位是秒，表示文件的
建立時間；接下來的 3 位元組是所在主機的唯一識別碼；接下來的 2 位元
組來自產生 ObjectId 的處理程序識別符號；後 3 位元組是一個自動增加的
計數器，確保相同處理程序同一秒產生的 ObjectId 是不一樣的。可以利用
ObjectId 的時間戳記對整個集合進行分段、快速檢查。當集合文件有上億
筆時，採用分頁查詢，越往後越慢。此時如果巧妙利用 ObjectId 的時間戳
記，可以將集合根據時間分段，多執行緒地去檢查集合。

```
1.  // 測試使用 mongoTemplate 尋找 name 是 " 張三 " 的記錄，列印主鍵 _id
2.  @Test
3.  @Order(10)
4.  fun testFindObjectId() {
5.      val studentCursor = mongoTemplate.db.getCollection("student").find()
6.              .filter(eq("name", " 張三 "))
7.              .cursor()
8.      while(studentCursor.hasNext()){
9.          val document = studentCursor.next()
10.         println("_id: ${document["_id"]}, date: ${document["date"]}")
11.     }
12.     studentCursor.close()
13. }
```

testDelete 方法使用 mongoTemplate 刪除集合的文件，使用 document 定義刪除的條件，將姓名是「張三」的記錄全部刪除。

```
1.  // 測試使用 mongoTemplate 刪除 name 是 " 張三 " 的所有記錄
2.  @Test
3.  @Order(11)
4.  fun testDelete() {
5.      val document = Document()
6.      document["name"] = " 張三 "
7.      mongoTemplate.db.getCollection("student").deleteMany(document)
8.      val n= mongoTemplate.db.getCollection("student").countDocuments()
9.      Assert.assertEquals(0, n)
10. }
```

4.5 Kotlin 整合 Spring Security

Spring Security 是為以 Spring 為基礎的企業應用系統提供宣告式的安全存取控制的解決方案。它利用 Spring 的依賴植入和針對切面技術，提供了宣告式的安全存取控制功能，可以在 Web 請求等級和方法呼叫等級處理身份認證和授權。本節主要介紹 Kotlin 整合 Spring Security 開發。

4.5.1 Spring Security 介紹

身份認證是指驗證某個使用者是否是系統中的合法使用者，一般要求使用者透過登入介面填寫使用者名稱、密碼及驗證碼等資訊。系統透過驗證使用者名稱和密碼完成認證。使用者授權是指某個使用者是否有許可權執行某個操作。系統會為不同的使用者分配不同的角色，每個角色對應一系列的許可權。Spring Security 在進行使用者認證以及授予許可權的時候，透過各式各樣的攔截器來控制許可權的存取，進一步實現系統安全存取介面。

Spring Security 的主要模組有以下幾個。

- ACL：支援透過存取控制清單為域物件提供安全性。
- Aspects：當使用 Spring Security 註釋時，會使用以 AspectJ 為基礎的切面。
- CAS Client：提供與 CAS 整合的功能。
- Configuration：包含對 XML、Java 設定功能的支援。
- Core：基本的功能函數庫。
- Cryptography：提供加密和解密編碼相關功能。
- LDAP：提供以 LDAP 為基礎的認證。
- OpenID：支援使用 OpenID 進行集中式認證。
- Web：提供以 Filter 為基礎的 Web 安全支援。

在使用者認證方面，Spring Security 支援主流的認證方式 ——HTTP 基本認證、HTTP 表單驗證、HTTP 摘要認證、OPENID 和 LDAP 等。Spring Security 支援以記憶體為基礎的驗證、以資料庫為基礎的驗證及使用者自訂服務的驗證。使用者一次完整的登入驗證和授權，是一個請求經過層層攔截器進一步實現許可權控制的過程。整個 Web 端設定為 DelegatingFilterProxy，它是所有篩檢程式鏈的代理類別，真正執行攔截處理的是由 Spring 容器管理的 Filter Bean 組成的 FilterChain。使用者認證涉及的主要類別有 UserDetails，其作為整個登入過程的 POJO 介面；UserDetailsService 用 於 產 生 UserDetails；UsernamePasswordAuthen ticationFilter，使用初始化中的 AuthenticationManager 的 AbstractUserDet ailsAuthenticationProvider 的 authenticate 方法進行驗證；AbstractUserDe tailsAuthenticationProvider，首先會透過 retrieveUser 方法使用我們定義的 UserDetailsService 產生對應的 UserDetails，獲得 UserDetails 後，再使用本身的 additionalAuthenticationChecks 方法去驗證資料庫的使用者資訊和登入資訊是否一致；如果 additionalAuthenticationChecks 沒有顯示出錯，那麼請求就會帶著 UserDetails 的許可權成功登入。

在使用者授權方面，Spring Security 提供以角色為基礎的存取控制和存取控制清單。Spring Security 有三種不同的安全註釋：附帶的 @Secured 註釋，@EnableGlobalMethodSecurity 中 設 定 securedEnabled = true；JSR-250 的 @RolesAllowed 註 釋，@EnableGlobalMethodSecurity 中設定 jsr250Enabled = true；運算式驅動的註釋。@PreAuthorize 方法呼叫之前，以運算式為基礎的計算結果限制對方法的存取；@PostAuthorize 方法呼叫之後，如果運算式的結果為 false，則拋出例外。@PreFilter 方法呼叫之前，過濾進入方法的輸入值；@PostFilter 方法呼叫之後，過濾方法的結果值。

Spring Security 在很多企業級應用中用於後台角色許可權控制、授權認證、安全防護（防止跨網站請求）、防止 Session 攻擊，非常容易在 Spring MVC 中使用。

4.5.2 使用 Kotlin 操作 Spring Security

Kotlin 整合 Spring Security 需要在 pom.xml 中增加以下依賴：

```
1.  <!— Spring Boot Security 相依套件 -->
2.  <dependency>
```

```
3.      <groupId>org.springframework.boot</groupId>
4.      <artifactId>spring-boot-starter-security</artifactId>
5.  </dependency>
```

WebSecurityConfig.kt 繼承自 WebSecurityConfigurerAdapter，多載 configure (HttpSecurity) 方法，設定如何透過攔截器保護請求。

```
1.  @Configuration
2.  @EnableWebSecurity// 啟用安全
3.  @EnableGlobalMethodSecurity(prePostEnabled = true)// 開啟註釋
4.  class WebSecurityConfig: WebSecurityConfigurerAdapter() {
5.      // 所有的介面都要進行授權才能存取，關閉跨域保護
6.      override fun configure(http: HttpSecurity) {
7.          http.formLogin()
8.                  .permitAll()
9.                  .and()
10.                 .authorizeRequests()
11.                 .antMatchers("/**")
12.                 .authenticated()
13.                 .and()
14.                 .csrf().disable()
15.     }
16. }
```

CustomAuthProvider.kt 繼承自 AuthenticationProvider，多載 authenticate (Authentication) 方法，對使用者許可權進行驗證。根據登入使用者名稱、密碼去資料庫查詢，如果能查詢到，則使用者登入成功，再根據使用者名稱尋找對應的許可權。

```
1.  @Component
2.  class CustomAuthProvider: AuthenticationProvider {
3.      // 植入 UserRepository
4.      @Autowired
5.      lateinit var userRepository: UserRepository
6.      // 使用者、角色列表
```

```
7.      private val auth = mapOf(Pair("test03", "ROLE_USER"), Pair("test02",
   "ROLE_ADMIN"))
8.      // 授權方法
9.      @Throws(AuthenticationException::class)
10.     override fun authenticate(authentication: Authentication):
   Authentication? {
11.         // 使用者名稱、密碼
12.         val username = authentication.name
13.         val password = authentication.credentials.toString()
14.         // 使用使用者名稱、密碼尋找是否有這個使用者
15.         val user = userRepository.findByUserName(username)
16.         // 取得該使用者的角色
17.         if(user?.password.equals(password)) {
18.             val authorities = AuthorityUtils.commaSeparatedString
   ToAuthorityList (auth[username])
19.             return UsernamePasswordAuthenticationToken(user, password,
   authorities)
20.         }
21.         return null
22.     }
23.     override fun supports(p0: Class<*>?): Boolean {
24.         return true;
25.     }
26. }
```

IndexController.kt 定義了幾個介面和對應的許可權驗證。

```
1.  @RestController
2.  class IndexController {
3.      // 測試介面，任何角色的使用者都可以存取
4.      @GetMapping("/")
5.      fun index(): String {
6.          return "Hello, Kotlin for Spring Boot!!"
7.      }
8.      // 測試介面，USER 角色的使用者可以存取，在存取方法前進行角色驗證
9.      @PreAuthorize("hasRole('USER')")
```

```kotlin
10.     @GetMapping("/hello/pre/user")
11.     fun rolePreUserHello(): String {
12.         println("pre filter admin user")
13.         return "Hello, I have admin role"
14.     }
15.     // 測試介面，USER 角色的使用者可以存取，在存取方法後進行角色驗證
16.     @PostAuthorize("hasRole('USER')")
17.     @GetMapping("/hello/post/user")
18.     fun rolePostUserHello(): String {
19.         println("post filter user role")
20.         return "Hello, I have user role"
21.     }
22.     // 測試介面，ADMIN 角色的使用者可以存取，在存取方法前進行角色驗證
23.     @PreAuthorize("hasRole('ADMIN')")
24.     @GetMapping("/hello/admin")
25.     fun roleAdminHello(): String {
26.         println("pre filter admin user")
27.         return "Hello, I have admin role"
28.     }
29.     // 測試介面，ADMIN、USER 角色的使用者可以存取，在存取方法前進行角色驗證
30.     @PreAuthorize("hasAnyRole('USER', 'ADMIN')")
31.     @GetMapping("/hello/any")
32.     fun anyRoleUserHello(): String {
33.         return "Hello, I have one of [user, admin] role"
34.     }
35.     // 測試介面，USER 角色的使用者可以存取，在存取方法前進行角色驗證
36.     // 只傳遞年齡大於 50 的 user 的資料
37.     @PreFilter(value="hasRole('USER') and filterObject.age > 50")
38.     @PostMapping("/user/prefilter")
39.     fun preFilterUser(@RequestBody user: List<User>): List<User> {
40.         println("pre filter user")
41.         return user
42.     }
43.     // 測試介面，USER 角色的使用者可以存取，在存取方法後進行角色驗證
44.     // 只傳遞年齡大於 50 的 user 的資料
```

```
45.     @PostFilter(value="hasRole('USER') and filterObject.age > 50")
46.     @PostMapping("/user/postfilter")
47.     fun postFilterUser(@RequestBody user: List<User>): List<User> {
48.         println("post filter user")
49.         return user
50.     }
51.     // 測試介面，ADMIN 角色的使用者可以存取，在存取方法前進行角色驗證
52.     // 只傳遞年齡大於 50 的 user 的資料
53.     @PreFilter(value="hasRole('ADMIN') and filterObject.age > 50")
54.     @PostMapping("/user/admin/prefilter")
55.     fun preFilterAdmin(@RequestBody user: List<User>): List<User> {
56.         println("pre filter user")
57.         return user
58.     }
59.     // 測試介面，USER 角色的使用者可以存取，在存取方法前進行角色驗證
60.     // 只傳遞 userName 等於 test02 的 user 的資料
61.     @PreFilter(value="hasRole('ADMIN') and filterObject.userName.equals
    ('test02')")
62.     @PostMapping("/user/prefilter1")
63.     fun preFilterAdmin1(@RequestBody user: List<User>): List<User> {
64.         println("pre filter user")
65.         return user
66.     }
67. }
```

輸入使用者名稱 test02、密碼 test02 登入後顯示 Hello, Kotlin for Spring Boot!!，test02 使用者具有 admin 角色。

rolePreUserHello 方法用 @preAuthorize 驗證使用者是否具有 "USER" 角色。rolePostUserHello 方法用 @postAuthorize 驗證使用者是否具有 "USER" 角色。@preAuthorize 在方法呼叫之前，以運算式為基礎的計算結果來限制對方法的存取；@PostAuthorize 在方法呼叫之後，如果運算式的結果為 false，則拋出例外。呼叫 rolePostUserHello 方法，雖然 test02 使用者沒有 "USER" 角色，仍然列印 "post filter user role"。

roleAdminHello 方法使用 @preAuthorize 驗證使用者是否具有 "ADMIN" 角色。anyRoleUserHello 方法驗證使用者是否具有 "ADMIN" 或 "USER" 角色。test02 使用者具有 "USER" 角色，會正常存取 anyRoleUserHello 方法，沒有許可權存取 roleAdminHello 方法。

preFilterUser 方法使用 @PreFilter 驗證使用者是否具有 "USER" 角色，並對輸入的參數進行過濾，把年齡大於 50 的 user 物件傳入。postFilterUser 方法使用 @PostFilter 驗證使用者是否具有 "USER" 角色，對傳回結果進行過濾。@PreFilter 在方法呼叫之前，過濾進入方法的輸入值；@PostFilter 在方法呼叫之後，過濾方法的結果值。兩個方法都傳回空，如圖 4.2 所示。

圖 4.2　/user/prefilter 介面傳回參數

preFilterAdmin1 方法使用 @PreFilter 驗證使用者是否具有 "ADMIN" 角色,並對輸入的參數進行過濾,把 userName 等於 "test02" 的 user 物件傳入,如圖 4.3 所示。

圖 4.3 /user/prefilter1 介面傳回參數

preFilterAdmin 方法使用 @ PreFilter 驗證使用者是否具有 "ADMIN" 角色，並對輸入的參數進行過濾，把年齡大於 50 的 user 物件傳入，如圖 4.4 所示。

圖 4.4　/admin/prefilter 介面傳回參數

4.6 Kotlin 整合 RocketMQ

RocketMQ 是一款用 Java 語言撰寫的分散式的佇列模型的開放原始碼訊息中介軟體，支援交易訊息、順序訊息、批次訊息、定時訊息及訊息回溯等。本節介紹 Kotlin 整合 RocketMQ 開發。

4.6.1 RocketMQ 介紹

RocketMQ 由阿里巴巴研發，現在是 Apache 頂級專案。其採用 Netty NIOl 架構實現資料通訊，用輕量級的 NameServer 進行網路路由，提供了服務效能，並支援訊息失敗重試機制。它支援叢集模式、消費者負載平衡、水平擴充能力並支援廣播模式。其採用零拷貝原理，順序寫碟、支援億級訊息堆積能力。

RocketMQ 架構分為四個模組：生產者，消費者，NameServer，Broker Server。元件均設計為無狀態，任何元件的節點都支援叢集部署擴充。

生產者充當訊息發行者的角色，支援分散式叢集方式部署。生產者透過 MQ 的負載平衡模組選擇對應的 Broker 叢集佇列進行訊息投遞，投遞的過程支援快速失敗並且低延遲。

消費者充當訊息消費者的角色，支援分散式叢集方式部署。支援以 push 和 pull 兩種模式對訊息進行消費。同時也支援叢集方式和廣播形式的消費，它提供即時訊息訂閱機制，可以滿足大多數使用者的需求。

NameServer 是以服務註冊發現功能為基礎的無狀態元件，支援獨立部署。NameServer 接收 Broker 叢集的註冊資訊並且將其儲存下來作為路由資訊的基本資料，然後提供心跳檢測機制，檢查 Broker 是否還存活。每個 NameServer 儲存關於 Broker 叢集的整個路由資訊和用於用戶端查詢的佇列資訊，然後生產者和消費者透過 NameServer 就可以知道整個 Broker 叢集的路由資訊，進一步進行訊息的投遞和消費。

Broker 是以高性能和低延遲為基礎的檔案儲存的無狀態元件，支援獨立部署。主要負責訊息的儲存、投遞和查詢以及保障服務的高可用性。Broker 包含以下模組。

- Remoting Module：整個 Broker 的實體，負責處理來自用戶端的請求。
- Client Manager：負責管理用戶端（生產者 / 消費者）和維護消費者的 topic 訂閱資訊。
- Store Service：提供簡單方便的 API 介面，提供將訊息儲存到物理硬碟和查詢的功能。
- HA Service：高可用服務，提供 master Broker 和 slave Broker 之間的資料同步功能。
- Index Service：根據特定的 Message key 對投遞到 Broker 的訊息進行索引服務，以提供訊息的快速查詢。

RocketMQ 支援叢集部署，可加強系統的傳輸量和可用性。namesrv 可以獨立部署，且 namesrv 與 namesrv 之間並無直接或間接的連結，雙方不存在心跳檢測，所以 namesrv 之間不存在主從切換過程，如果其中一台 namesrv 當機，生產者和消費者會直接從另一台 namesrv 中請求資料。Broker 支援主從架構模式，目前它支援 master 寫入操作，只有當 master 讀取壓力高於某個點（master 訊息拉取出現堆積）時，才會將讀取壓力轉給 slaver。無法做到主從切換，master 當機，slaver 只能提供訊息消費，slaver 不會被選舉為 master 來繼續工作。如果 master 當機，整個訊息佇列環境近乎癱瘓。如果使用者體量稍微大一些，單 master/ 單 slaver 架構扛不住，可以採用多 master/ 多 slaver 部署架構。

RocketMQ 提供了兩種寫入磁碟方式，即同步寫入磁碟和非同步寫入磁碟。同步寫入磁碟指訊息投放到 Broker 之後，會在寫入檔案之後才傳回成功，同步寫入磁碟可保障資料不遺失。非同步寫入磁碟則指訊息投放到 Broker 成功後即可傳回，同時啟動另外的執行緒來儲存訊息。

RocketMQ 可以應用於非同步解耦、訊息的順序收發、削峰填谷、分散式交易資料的一致性及大規模機器的快取同步等業務場景。

4.6.2 使用 Kotlin 操作 RocketMQ

Kotlin 整合 RocketMQ 需要在 pom.xml 中增加以下依賴：

```
1.  <!-- Spring Boot RocketMQ 相依套件 -->
2.  <dependency>
3.      <groupId>org.apache.rocketmq</groupId>
4.      <artifactId>rocketmq-spring-boot-starter</artifactId>
5.      <version>2.0.4</version>
6.  </dependency>
```

在 application.yml 中增加以下 RocketMQ 的設定，我們使用的是單機 RocketMQ：

```
1.  rocketmq:
2.    name-server: 127.0.0.1:9876                      # RocketMQ 的主機、通訊埠
3.    producer:
4.      group: kotlin-group                            # 生產者 group
5.      send-message-timeout: 300000                   # 發送訊息的逾時
6.      compress-message-body-threshold: 4096          # 壓縮訊息的設定值 4KB
7.      max-message-size: 4194304                      # 訊息最大 4MB
8.      retry-times-when-send-failed: 2                # 發送失敗後重試 2 次
9.      retry-next-server: true                        # 開啟內部訊息重試
10.     retry-times-when-send-async-failed: 0          # 非同步發送重試 0 次
```

OrderPaidEvent.kt 定義了一個訊息實體：

```
1.  // 訊息實體，具有兩個屬性 orderId 和 paidMoney
2.  class OrderPaidEvent(
```

```
3.        val orderId: String,
4.        val paidMoney: BigDecimal
5. ){
6.    constructor():this("0", BigDecimal(0.0))
7.    override fun toString(): String {
8.        return "OrderPaidEvent($orderId, $paidMoney)"
9.    }
10. }
```

MqProducer.kt 定義訊息的生產者：

```
1. @Component
2. class MqProducer {
3.     // 植入 rocketMQTemplate
4.     @Autowired
5.     lateinit var rocketMQTemplate: RocketMQTemplate
6.     // 向 "kotlin-topic" 發送一筆訊息
7.     fun sendMessage(orderPaidEvent: OrderPaidEvent) {
8.         println("send message: $orderPaidEvent")
9.         rocketMQTemplate.send("kotlin-topic", MessageBuilder.withPayload
   (orderPaidEvent).build())
10.     }
11.     // 向 "kotlin-topic" 發送一筆訊息，tag 是 kotlin-tag
12.     fun sendMessageWithTag(orderPaidEvent: OrderPaidEvent) {
13.         println("send message: $orderPaidEvent, tag: kotlin-tag")
14.         rocketMQTemplate.convertAndSend("kotlin-topic:kotlin-tag",
   orderPaidEvent)
15.     }
16.     // 使用 convertAndSend 方法向 "kotlin-topic" 發送一筆訊息
17.     fun convertAndSendMessage(orderPaidEvent: OrderPaidEvent) {
18.         println("convertAndSend message: $orderPaidEvent")
19.         rocketMQTemplate.convertAndSend("kotlin-topic", orderPaidEvent)
20.     }
21.     // 向 "kotlin-topic" 非同步發送一筆訊息，發送成功後執行 SendCallback
   回呼函數
22.     fun asyncSendMessage(orderPaidEvent: OrderPaidEvent) {
```

```
23.          println("async send single message: $orderPaidEvent")
24.          rocketMQTemplate.asyncSend("kotlin-topic", orderPaidEvent,
    object: SendCallback{
25.              override fun onSuccess(p0: SendResult?) {
26.                  println("async send success: $p0")
27.              }
28.              override fun onException(p0: Throwable?) {
29.                  throw Exception(p0)
30.              }
31.          },1000L)
32.      }
33.      // 向 "kotlin-topic" 批次同步發送多筆訊息，逾時為 60 秒
34.      fun syncSendBatchMessage(orderPaidEvents: List<OrderPaidEvent>) {
35.          println("sync send single message: $orderPaidEvents")
36.          val msg = orderPaidEvents.map { o -> MessageBuilder.
    withPayload(o).build() }
37.          rocketMQTemplate.syncSend("kotlin-topic", msg, 60000L)
38.      }
39.      // 向 "kotlin-topic" 同步發送一筆訊息
40.      fun syncSendMessage(orderPaidEvent: OrderPaidEvent) {
41.          println("sync send single message: $orderPaidEvent")
42.          rocketMQTemplate.syncSend("kotlin-topic", orderPaidEvent)
43.      }
44. }
```

MqConsumer.kt 定義訊息的消費者：

```
1.  @Component
2.  // 開啟 @RocketMQ 訊息監聽註釋，消費 kotlin-topic 的訊息，消費群組是 kotlin-
    consumer
3.  @RocketMQMessageListener(topic = "kotlin-topic", consumerGroup = "kotlin-
    consumer")
4.  class MqConsumer: RocketMQListener<OrderPaidEvent> {
5.      // 消費訊息後列印
6.      override fun onMessage(p0: OrderPaidEvent?) {
7.          println("OrderPaidEventConsumer received: $p0")
```

```
8.     }
9.  }
```

MqTagConsumer.kt 定義消費的訊息的 tag 是 "kotlin-tag"：

```
1.  @Component
2.  // 開啟 @RocketMQ 訊息監聽註釋，消費 kotlin-topic 的訊息，消費群組是 kotlin-
    consumer1
3.  // 只消費 tag 是 kotlin-tag 的訊息
4.  @RocketMQMessageListener(topic = "kotlin-topic", consumerGroup = "kotlin-
    consumer1", selectorExpression = "kotlin-tag")
5.  class MqTagConsumer: RocketMQListener<OrderPaidEvent> {
6.      // 消費訊息後列印
7.      override fun onMessage(p0: OrderPaidEvent?) {
8.          println("OrderPaidEventConsumer received: $p0")
9.      }
10. }
```

MqController.kt 定義了幾個介面測試 RocketMQ：

```
1.  @RestController
2.  class MqController {
3.      @Autowired
4.      lateinit var mqProducer: MqProducer
5.      // 測試介面，測試發送單筆訊息
6.      @PostMapping("/mq/send")
7.      fun sendMsg(@RequestBody orderPaidEvent: OrderPaidEvent) {
8.          mqProducer.sendMessage(orderPaidEvent)
9.      }
10.     // 測試介面，測試發送單筆訊息，並打上標籤
11.     @PostMapping("/mq/send/tag")
12.     fun sendMsgTag(@RequestBody orderPaidEvent: OrderPaidEvent) {
13.         mqProducer.sendMessageWithTag(orderPaidEvent)
14.     }
15.     // 測試介面，測試使用 convertAndSendMessage 方法發送單筆訊息
16.     @PostMapping("/mq/convertAndSend")
```

```
17.    fun convertAndSendMsg(@RequestBody orderPaidEvent: OrderPaidEvent) {
18.        mqProducer.convertAndSendMessage(orderPaidEvent)
19.    }
20.    // 測試介面，測試非同步發送單筆訊息
21.    @PostMapping("/mq/asyncSend")
22.    fun asyncAndSendMsg(@RequestBody orderPaidEvent: OrderPaidEvent) {
23.        mqProducer.asyncSendMessage(orderPaidEvent)
24.    }
25.    // 測試介面，測試批次同步發送訊息
26.    @PostMapping("/mq/asyncBatchSend")
27.    fun asyncAndBatchSendMsg(@RequestBody orderPaidEvents: List
   <OrderPaidEvent>) {
28.        mqProducer.syncSendBatchMessage(orderPaidEvents)
29.    }
30.    // 測試介面，測試同步發送單筆訊息
31.    @PostMapping("/mq/syncSend")
32.    fun syncSendMsg(@RequestBody orderPaidEvent: OrderPaidEvent) {
33.        mqProducer.syncSendMessage(orderPaidEvent)
34.    }
35. }
```

sendMsg 方法向 topic："kotlin-topic" 發送一筆訊息 orderPaidEvent，MqConsumer 會消費這筆訊息。sendMsgTag 方法發送一分散連結有 tag 標籤的訊息，tag 是 "kotlin-tag"，MqConsumer 和 MqTagConsumer 都會消費這筆訊息，但是 MqTagConsumer 只會消費 tag 是 "kotlin-tag" 的訊息。convertAndSendMsg 方法用 convertAndSend 方法發送了一筆訊息，convertAndSend 方法會對訊息實體進行包裝，然後發送。asyncAndSendMsg 非同步發送一筆訊息，並定義了回呼方法 SendCallBack，訊息發送成功會執行該方法。asyncAndBatchSendMsg 非同步批次發送訊息，如圖 4.5 所示，批次發送三筆訊息。syncSendMsg 同步發送一筆訊息。

圖 4.5　訊息佇列批次發送訊息介面參數

使用 RocketMQ-Console-Ng 可以檢視 "kotlin-topic" 的消費情況。kotlin-topic 有 4 個佇列，其中佇列 0、2、3 各發送了 2、2、1 筆訊息，broker Offset 和 consumerOffset 相等，沒有出現消費落後，如圖 4.6 所示。

Topic	kotlin-topic		Delay	0	LastConsumeTime		2020-01-29 10:59:06	

broker	queue	consumerClient	brokerOffset	consumerOffset	diffTotal	lastTimestamp
DESKTOP-644KAN2	0		2	2	0	2020-01-29 10:59:06
DESKTOP-644KAN2	1		0	0	0	1970-01-01 08:00:00
DESKTOP-644KAN2	2		2	2	0	2020-01-29 10:58:46
DESKTOP-644KAN2	3		1	1	0	2020-01-29 10:58:56

圖 4.6　kotlin-topic 消費情況

4.7 Kotlin 整合 Elasticsearch

Elasticsearch 是一個分散式的開放原始碼搜尋和分析引擎，適用於所有類型的資料，包含文字、數字、地理空間、結構化及非結構化的資料。Elasticsearch 是在 Apache Lucene 的基礎上開發而成的，於 2010 年發佈。本節介紹使用 Kotlin 整合 Elasticsearch 進行開發。

4.7.1 Elasticsearch 介紹

Elasticsearch 具有簡單的 REST 風格 API，兼具分散式特性和高可擴充性，是 Elastic Stack 的核心元件。Elastic Stack 是適用於資料獲取、儲存、分析、視覺化的一組開放原始碼工具，包含 Elasticsearch、Logstash、Kibana 等。Elastic Stack 簡化了資料獲取、視覺化和報告過程，透過與 Beats 和 Logstash 進行整合，使用者能夠在 Elasticsearch 中索引資料之前輕鬆地處理資料。Kibana 不僅可針對 Elasticsearch 資料提供即時視覺化，同時還提供 UI 以便使用者快速存取應用程式效能監測（APM）、記錄檔和基礎設施指標等資料。

原始資料會從多個來源（包含記錄檔、系統指標和網路應用程式）被輸入到 Elasticsearch 中。資料獲取是在 Elasticsearch 中進行索引之前解析、標準化並充實這些原始資料的過程。這些資料在 Elasticsearch 中索引完成之後，使用者便可針對資料進行複雜的查詢，並使用聚合來檢索本身資料的複雜整理。

Elasticsearch 索引指相互連結的文件集合。Elasticsearch 會以 JSON 文件的形式儲存資料。每個文件都會在一組鍵（欄位或屬性的名稱）和它們對應的值（字串、數字、布林值、日期、數值群組、地理位置或其他類型的資料）之間建立聯繫。

Elasticsearch 使用的是一種被稱為倒排索引的資料結構，這種結構允許快速地進行全文字搜尋。倒排索引會列出在所有文件中出現的每個特有詞彙，並且可以找到包含每個詞彙的全部文件。

在索引過程中，Elasticsearch 會儲存文件並建置倒排索引，這樣使用者便可以近即時地對文件資料進行搜尋。索引過程是在索引 API 中啟動的，透過此 API，你既可在特定索引中增加 JSON 文件，也可更改特定索引中的 JSON 文件。

Elasticsearch 的檢索速度很快。由於 Elasticsearch 是在 Lucene 基礎上建置而成的，所以在全文字搜尋方面表現十分出色。Elasticsearch 同時還是一個近即時的搜尋平台，這表示從文件索引操作到文件變為可搜尋狀態之間的延遲很短，一般只有一秒。因此，Elasticsearch 非常適用於對時間有嚴苛要求的使用案例，舉例來說，安全分析和基礎設施監測。

Elasticsearch 具有分散式的本質特徵。Elasticsearch 中儲存的文件分佈在不同的容器中，這些容器被稱為分片，可以進行複製以提供資料容錯備份，以防發生硬體故障。Elasticsearch 的分散式特徵使得它可以擴充至數百台（甚至數千台）伺服器，並處理 PB 量級的資料。

Elasticsearch 還包含一系列廣泛的功能。除了速度、可擴充性和彈性等優勢以外，它還有大量強大的內建功能（舉例來說，資料整理和索引生命週期管理），可以方便使用者更加高效率地儲存和搜尋資料。

Elasticsearch 支援多種程式語言，官方提供了針對 Java、JavaScript（Node.js）、Go、.NET（C#）、PHP、Perl、Python、Ruby 語言的用戶端。Elasticsearch 提供強大且全面的 REST API 集合，這些 API 可用來執行各種任務，舉例來說，檢查叢集的執行狀況、針對索引執行 CRUD（建立、讀取、更新、刪除）和搜尋操作，以及執行諸如篩選和聚合等進階搜尋操作。

Elasticsearch 能夠索引多種類型的內容，可應用於應用程式搜尋、網站搜尋、企業搜尋、記錄檔處理和分析、基礎設施指標和容器監測、應用程式效能監測、地理空間資料分析、視覺化、安全分析及業務分析等場景。

4.7.2 使用 Kotlin 操作 Elasticsearch

Kotlin 整合 Elasticsearch 需要在 pom.xml 中增加以下依賴：

```
1.  <!-- Spring Boot Elasticsearch 相依套件 -->
2.  <dependency>
3.      <groupId>org.springframework.boot</groupId>
4.      <artifactId>spring-boot-starter-data-elasticsearch</artifactId>
5.  </dependency>
```

在 application.yml 中增加以下設定，設定 Elasticsearch 的叢集，我們使用的是單機 Elasticsearch：

```
1.  spring:
2.    data:
3.      elasticsearch:
4.        cluster-name: elasticsearch        # 叢集名稱
5.        cluster-nodes: 127.0.0.1:9300      # elasticsearch host port
```

Item.kt 定義了 Elasticsearch 中的實體物件：

```
1.  // 定義實體 Item
2.  @Document(indexName = "item",type = "docs", shards = 1, replicas = 0)
3.  data class Item(
4.      @Id
5.      val id: Long,
6.      // 對 title 使用 ik_max_word 進行分詞
7.      @Field(type = FieldType.Text, analyzer = "ik_max_word")
8.      val title: String,
9.      @Field(type = FieldType.Keyword)
10.     val category: String,
```

```
11.        @Field(type = FieldType.Keyword)
12.        val brand: String,
13.        @Field(type = FieldType.Double)
14.        val price: Double,
15.        @Field(index = false, type = FieldType.Keyword)
16.        val images: String
17. ){
18.        constructor():
19.                this(0L, "", "", "", 0.0, "")
20. }
```

ItemRepository.kt 定義了 Item 集合的 CRUD 操作：

```
1.  interface ItemRepository: ElasticsearchRepository<Item, Long>
```

ItemRepositoryTest.kt 對 Item 集合的常見操作進行了測試。createIndex、putMapping 方法分別建立了 Item 索引和對映：

```
1.  @SpringBootTest
2.  @TestMethodOrder(MethodOrderer.OrderAnnotation::class)
3.  @ExtendWith(SpringExtension::class)
4.  class ItemRepositoryTest {
5.      @Autowired
6.      lateinit var elasticsearchTemplate: ElasticsearchTemplate
7.
8.      @Autowired
9.      lateinit var itemRepository: ItemRepository
10.     // 建立索引、對映
11.     @Test
12.     @Order(1)
13.     fun testCreateIndex() {
14.         elasticsearchTemplate.createIndex(Item::class.java)
15.         elasticsearchTemplate.putMapping(Item::class.java)
16.     }
17. }
```

儲存一筆新的 Item 記錄：

```
1.  // 測試插入一筆 Item 記錄
2.  @Test
3.  @Order(2)
4.  fun testAddNewItem() {
5.      val item = Item(1L, "Iphone 11", " 手機 ", " 蘋果 ", 5899.0, "http://
    image.baidu.com/13123.jpg")
6.      itemRepository.save(item)
7.  }
```

儲存多筆 Item 記錄：

```
1.  // 測試批次插入 Item 記錄
2.  @Test
3.  @Order(3)
4.  fun testBatchAddNewItems(){
5.      val items = arrayOf(
6.          Item(2L, " 堅果手機 R1", " 手機 ", " 錘子 ", 3699.00, "http://
    image.baidu.com/13123.jpg"),
7.          Item(3L, " 華為 MATE10", " 手機 ", " 華為 ", 4499.00, "http://
    image.baidu.com/13123.jpg")
8.      )
9.      itemRepository.saveAll(items.asList())
10. }
11.
```

更新 id=1 的 Item 記錄：

```
1.  // 測試更新 Item 記錄
2.  @Test
3.  @Order(4)
4.  fun testUpdateItem(){
5.      val item = Item(1L, " 蘋果 XS Max", " 手機 ", " 蘋果 ", 4899.00,
    "http://image.baidu.com/13123.jpg")
6.      itemRepository.save(item)
7.  }
```

尋找所有的 Item 記錄：

```
1.  // 測試尋找 Item 集合中的所有記錄
2.  @Test
3.  @Order(5)
4.  fun testFindAll(){
5.      val items = itemRepository.findAll()
6.      Assert.assertEquals(3, items.toList().size)
7.  }
```

分頁尋找 Item 記錄，每頁傳回 2 筆記錄，尋找第 1 頁的記錄：

```
1.  // 測試分頁尋找 Item 記錄
2.  @Test
3.  @Order(6)
4.  fun testFindByPage(){
5.      val items = itemRepository.findAll(PageRequest.of(1, 2))
6.      items.forEach { println(it) }
7.  }
```

尋找所有 Item 記錄，並根據價格降冪排列：

```
1.  // 測試根目錄據 price 降冪尋找 Item 記錄
2.  @Test
3.  @Order(7)
4.  fun testFindBySort(){
5.      val items = itemRepository.findAll(Sort.by("price").descending())
6.      items.forEach { println(it) }
7.  }
```

尋找 title 是「堅果手機」的 Item 記錄，matchQuery 進行詞條比對，先分詞然後再查詢結果：

```
1.  // 測試尋找 title 包含 "堅果手機" 的 Item 記錄，進行分詞
2.  @Test
3.  @Order(8)
4.  fun testMatchQuery(){
```

```
5.      val queryBuilder = NativeSearchQueryBuilder()
6.      queryBuilder.withQuery(QueryBuilders.matchQuery("title", "堅果手機"))
7.      val items = itemRepository.search(queryBuilder.build())
8.      Assert.assertEquals(1, items.totalElements)
9.      items.forEach { println(it) }
10. }
```

尋找 title 是「堅果」的 Item 記錄，使用 termQuery 進行詞條比對，不分詞：

```
1.  // 測試尋找 title 包含 "堅果" 的 Item 記錄，不分詞
2.  @Test
3.  @Order(9)
4.  fun testTermQuery(){
5.      val queryBuilder = NativeSearchQueryBuilder()
6.      queryBuilder.withQuery(QueryBuilders.termQuery("title", "堅果"))
7.      val items = itemRepository.search(queryBuilder.build())
8.      Assert.assertEquals(1, items.totalElements)
9.      items.forEach { println(it) }
10. }
```

尋找 title 包含「堅果」的 Item 記錄，使用 fuzzyQuery 進行模糊查詢：

```
1.  // 測試尋找 title 包含 "堅果" 的 Item 記錄，模糊查詢
2.  @Test
3.  @Order(10)
4.  fun testFuzzyQuery(){
5.      val queryBuilder = NativeSearchQueryBuilder()
6.      queryBuilder.withQuery(QueryBuilders.fuzzyQuery("title", "堅果"))
7.      val items = itemRepository.search(queryBuilder.build())
8.      Assert.assertEquals(1, items.totalElements)
9.      items.forEach { println(it) }
10. }
```

尋找 title 是「堅果」、brand 是「錘子」的 Item 記錄。使用 booleanQuery，
布林查詢可查詢布林關係，包含 BooleanClause.Occur.MUST、Boolean

Clause.Occur.MUST_NOT、BooleanClause.Occur.SHOULD，分別表示必須包含、不能包含和可以包含三種。

```
1.  // 測試尋找 title 包含 " 堅果 "，brand 是 " 錘子 " 的 Item 記錄
2.  @Test
3.  @Order(11)
4.  fun testBooleanQuery(){
5.      val queryBuilder = NativeSearchQueryBuilder()
6.      queryBuilder.withQuery(QueryBuilders.boolQuery()
7.              .must(QueryBuilders.termQuery("title", " 堅果 "))
8.              .must(QueryBuilders.termQuery("brand", " 錘子 ")))
9.
10.     val items = itemRepository.search(queryBuilder.build())
11.     Assert.assertEquals(1, items.totalElements)
12.     items.forEach { println(it) }
13. }
```

尋找價格在 3000 元到 4000 元的 Item 記錄，使用 rangeQuery 可進行範圍尋找：

```
1.  // 測試尋找價格在 3000 元到 4000 元之間的 Item 記錄
2.  @Test
3.  @Order(12)
4.  fun testRangeQuery(){
5.      val queryBuilder = NativeSearchQueryBuilder()
6.      queryBuilder.withQuery(QueryBuilders.rangeQuery("price").from(3000).
    to(4000))
7.
8.      val items = itemRepository.search(queryBuilder.build())
9.      Assert.assertEquals(2, items.totalElements)
10.     items.forEach { println(it) }
11. }
```

刪除 id=1 的 Item 記錄：

```
1.  // 測試刪除 Item 記錄
```

```
2.  @Test
3.  @Order(13)
4.  fun testDelete(){
5.      itemRepository.deleteById(1)
6.
7.      Assert.assertEquals(2, itemRepository.count())
8.  }
```

4.8 Kotlin 整合 Swagger

Swagger 是一個標準和完整的架構，用於產生、描述、呼叫和視覺化 RESTful 風格的 Web 服務。本節介紹使用 Kotlin 整合 Swagger 開發。

4.8.1 Swagger 介紹

Swagger 提供了建置 API 的一套工具和標準，按照它的標準去定義介面及介面相關的資訊，再透過 Swagger 衍生出來的一系列專案和工具，就可以產生各種格式的介面文件，產生多種語言的用戶端和服務端的程式，以及產生介面偵錯頁面等。在開發新版本或反覆運算版本的時候，只需更新 Swagger 描述檔案，就可以自動產生介面文件、用戶端及服務端程式，做到呼叫端程式、服務端程式以及介面文件的一致性。

Swagger 包含以下開放原始碼專案。

- Swagger Codegen：透過 Codegen 可以將描述檔案產生 HTML 格式或 cwiki 形式的介面文件，同時也能產生多種語言的服務端和用戶端的程式。支援透過 Jar 套件、Docker、Node.js 等方式在當地語系化執行產生，也可以在後面介紹的 Swagger Editor 中線上產生。
- Swagger UI：提供了一個視覺化的 UI 頁面展示描述檔案。介面的呼叫方、測試方、專案經理等都可以在該頁面中對相關介面進行查閱和做

一些簡單的介面請求。該專案支援線上匯入描述檔案和在本機部署 UI
專案。

- Swagger Editor：編輯 Swagger 描述檔案的編輯器，該編輯器支援即時
  預覽描述檔案的更新效果，還提供了線上編輯器和本機部署編輯器兩種
  方式。
- Swagger Inspector：可以對介面進行線上測試，比在 Swagger UI 中做介
  面請求，會傳回更多的資訊，也會儲存使用者的實際請求參數等資料。
- Swagger Hub：整合了上面所有專案的各項功能，以專案和版本為單位，
  將使用者的描述檔案上傳到 Swagger Hub 中。在 Swagger Hub 中可以完
  成上面介紹的專案的所有工作，需要註冊帳號，有免費版和收費版。

維護這個 JSON 或 YAML 格式的描述檔案有一定的工作量，在持續反覆運
算開發的時候，常常會忽略更新這個描述檔案，導致基於該描述檔案產生
的介面文件失去了參考意義。Springfox Swagger 基於 Swagger 標準，可以
自動檢查類別、控制器、方法、模型類別以及它們對映到的 URL，自動產
生 JSON 格式的描述檔案 進而產生與程式一致的介面文件和用戶端程式。

Swagger 常用的註釋有 @Api()，用於類別，表示這個類別是 Swagger
的資源；@ApiOperation()，用於方法，表示一個 HTTP 請求的操作；
@ApiParam()，用於方法、參數和欄位說明，表示對參數增加中繼資
料（說明或是否必填等）；@ApiModel()，用於類別，對類別說明；
@ApiModelProperty()，用於方法、欄位，對屬性欄位說明或對資料操作更
改說明；@ApiIgnore()，用於類別、方法、方法參數，表示這個方法或類
別被忽略；@ApiImplicitParam()，用於方法，表示這是單獨的請求參數；
@ApiImplicitParams() 用於方法，包含多個 @ApiImplicitParam。

對於前後端分離開發的場景，如果能在提供介面文件的同時，把所有介面
的模擬請求回應資料也提供給前端，或有 Mock 系統，直接將這些模擬資
料輸入 Mock 系統，那將加強前端的開發效率，減少許多發生在聯調時才

會發生的問題。透過適當地在程式中加入 Swagger 的註釋,可以讓介面文件描述資訊更加詳細,如果把每個出入參數的範例值都配上,那前端就可以直接在介面文件中拿到模擬資料。

4.8.2 使用 Kotlin 操作 Swagger

Kotlin 整合 Swagger 需要在 pom.xml 中增加以下依賴:

```
1.  <-- Swagger 相依套件 -->
2.  <dependency>
3.      <groupId>io.springfox</groupId>
4.      <artifactId>springfox-swagger2</artifactId>
5.      <version>2.9.2</version>
6.  </dependency>
7.  <dependency>
8.      <groupId>io.springfox</groupId>
9.      <artifactId>springfox-swagger-ui</artifactId>
10.     <version>2.9.2</version>
11. </dependency>
```

SwaggerConfig.kt 定義了介面的基本資訊:

```
1.  @Configuration
2.  @EnableSwagger2
3.  class SwaggerConfig {
4.      // 設定 apiInfo,定義套件路徑,掃描該路徑下的所有介面
5.      @Bean
6.      fun createRestApi(): Docket {
7.          return Docket(DocumentationType.SWAGGER_2)
8.                  .apiInfo(apiInfo())
9.                  .select()
10.                 .apis(RequestHandlerSelectors.basePackage ("io.kang.
    example.controller.swagger"))
11.                 .paths(PathSelectors.any())
12.                 .build()
```

```
13.      }
14.      // 設定 apiInfo 的實際資訊：標題、連絡人、版本編號、描述
15.      @Bean
16.      fun apiInfo(): ApiInfo {
17.          return ApiInfoBuilder()
18.                  .title(" 使用 Swagger2 建置 RESTful APIs")
19.                  .contact(Contact("dutyk", "https://github.com/dutyk",
     "1013812851@qq.com"))
20.                  .version("1.0")
21.                  .description("Demo for book Kotlin Spring Boot Action")
22.                  .build()
23.      }
24. }
```

SwaggerController.kt 定義了 get、post、put、delete 介面，分別對應 Get
Mapping、PostMapping、PutMapping 及 DeleteMapping，並用 @ApiOperation()
增加了對介面的描述，用 @ApiImplicitParam() 增加了對參數的描述。

```
1.  @RestController
2.  @RequestMapping("/users")
3.  class SwaggerController {
4.      @Autowired
5.      lateinit var userRepository: UserRepository
6.      // get 介面，取得使用者清單
7.      @ApiOperation(value=" 取得使用者列表 ", notes="")
8.      @GetMapping("/all")
9.      fun getUserList(): List<User> {
10.         return userRepository.findAll().toList()
11.     }
12.     // post 介面，新增使用者
13.     @ApiOperation(value=" 建立使用者 ", notes=" 根據 User 物件建立使用者 ")
14.     @ApiImplicitParam(name = "user", value = " 使用者詳細實體 user",
     required = true, dataType = "User")
15.     @PostMapping("/add")
16.     fun postUser(@RequestBody user: User): String {
```

```
17.         userRepository.save(user)
18.         return "success"
19.     }
20.     // get 介面，在介面傳遞參數：id，取得指定使用者資訊
21.     @ApiOperation(value=" 取得使用者詳細資訊 ", notes=" 根據 url 的 id 來取得
    使用者詳細資訊 ")
22.     @ApiImplicitParam(name = "id", value = " 使用者 ID", required = true,
    dataType = "Long")
23.     @GetMapping("/find/{id}")
24.     fun getUser(@PathVariable id: Long): User? {
25.         return userRepository.findById(id).get()
26.     }
27.     // put 介面，更新指定使用者的資訊
28.     @ApiOperation(value=" 更新使用者詳細資訊 ", notes=" 根據 url 的 id 來指定
    更新物件，並根據傳過來的 user 資訊來更新使用者詳細資訊 ")
29.     @ApiImplicitParams(
30.         ApiImplicitParam(name = "id", value = " 使用者 ID", required =
    true, dataType = "Long"),
31.         ApiImplicitParam(name = "user", value = " 使用者詳細實體 user",
    required = true, dataType = "User")
32.     )
33.     @PutMapping("/update/{id}")
34.     fun updateUser(@PathVariable id: Long, @RequestBody user: User) {
35.         val user = User(id, user.userName, user.password, user.email,
    user.age, user.height, user.address, user.education, user.income)
36.         userRepository.save(user)
37.     }
38.     // delete 介面，刪除指定使用者
39.     @ApiOperation(value=" 刪除使用者 ", notes=" 根據 url 的 id 來指定刪除物件 ")
40.     @ApiImplicitParam(name = "id", value = " 使用者 ID", required = true,
    dataType = "Long")
41.     @DeleteMapping("/delete/{id}")
42.     fun deleteUser(@PathVariable id: Long) {
43.         userRepository.deleteById(id)
44.     }
```

```
45. }
```

在瀏覽器中輸入 swagger-ui 的位址，可跳躍到介面偵錯頁面對定義的介面
進行測試。

「取得使用者清單」介面測試如下：點擊 "Execute" 可以測試這個介面，傳
回使用者清單，這個測試的是 get 請求，如圖 4.7 所示。

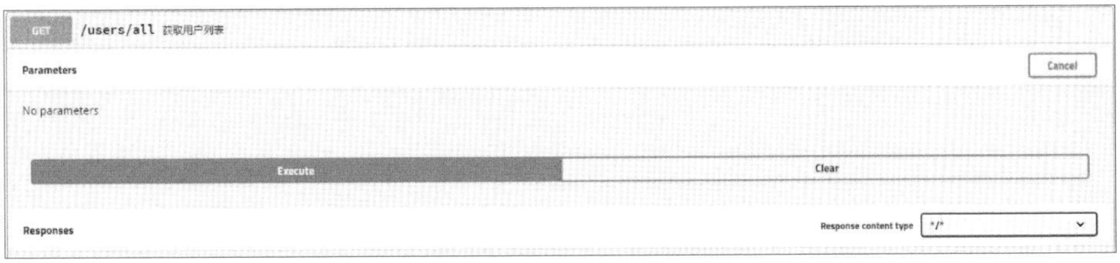

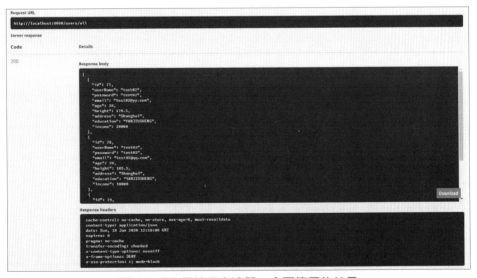

圖 4.7　取得使用者清單介面

介面的傳回結果如圖 4.8 所示。

圖 4.8「取得使用者清單」介面傳回的結果

「建立使用者」介面測試如下：輸入使用者的各個屬性的值，提交即可。

這個測試的是 post 請求，如圖 4.9 所示。

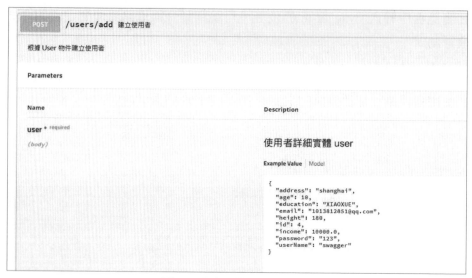

圖 4.9「建立使用者」介面

「根據 id 尋找使用者」介面測試如圖 4.10 所示，在輸入框中輸入使用者 id，這個測試的是 get 請求。

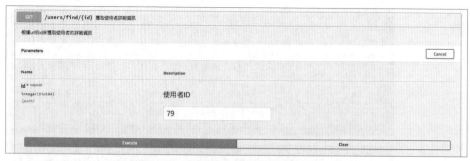

圖 4.10「根據 id 尋找使用者」介面

介面傳回 id 是 79 的使用者記錄，如圖 4.11 所示。

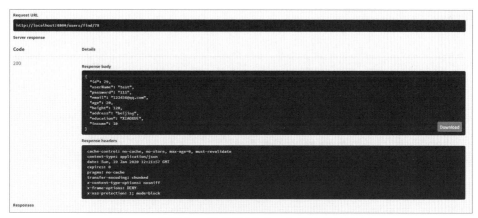

圖 4.11「根據 id 尋找使用者」介面傳回的記錄

「更新指定 id 使用者資訊」介面測試如下：在輸入框中輸入使用者的 id 為 79，並填寫使用者的資訊。這個測試的是 put 請求，如圖 4.12 所示。

圖 4.12「更新指定 id 使用者資訊」介面

「根據 id 刪除使用者」介面測試如下：輸入使用者 id=79，可以把 id=79 的記錄刪除，此時，查不到 id=79 的使用者記錄 3。這個測試的是 delete 請求，如圖 4.13 所示。

圖 4.13「根據 id 刪除使用者」介面

4.9 小結

本章介紹了使用 Kotlin 操作常用的微服務元件的方法，包含的元件有：
Spring Boot、Redis、JPA、QueryDSL、MongoDB、Spring Security、
RocketMQ、Elasticsearch、Swagger。Kotlin 可以使用已有的 Jar 套件，節
省了開發成本。本章對每個元件都提供了許多範例，方便大家在使用時參
考。

Kotlin 應用於微服務
註冊中心

微 服務註冊中心對於微服務系統很重要，有助系統解耦，開發分散式
應用系統。本章介紹將 Kotlin 應用於微服務註冊中心的相關知識，
主要包含四種常用的註冊中心：Eureka、Consul、Zookeeper、Nacos。

5.1 Eureka

Eureka 是 Netflix 公司開發的服務發現架構，是一個以 REST 為基礎的服
務。它主要以 AWS 雲端服務為支撐，提供服務發現並實現負載平衡和容
錯移轉。Spring Cloud 將它整合在其子專案 spring-cloud-netflix 中，以實現
Spring Cloud 的服務發現功能。本節介紹使用 Eureka 作為 Kotlin 開發的微
服務註冊中心的相關知識。

5.1.1 Eureka 介紹

Eureka 有 Server 和 Client 兩個元件。服務端提供服務註冊，當用戶端服務
啟動的時候，會主動向服務端進行註冊，服務端會儲存所有已經註冊的服

務節點的資訊。服務端會管理這些節點資訊，並且會將異常的節點從服務清單中移除。用戶端有快取功能，所以即使 Eureka 叢集中的所有節點都故障，或發生網路磁碟分割故障導致用戶端不能存取任何一台 Eureka 伺服器，Eureka 服務的消費者仍然可以透過 Eureka 用戶端快取來取得現有的服務註冊資訊。無論是服務端還是用戶端，都支援叢集模式，註冊資訊和更新資訊會在整個 Eureka 叢集的節點中進行複製。

Eureka Client 是一個 Java 用戶端，用於簡化與 Eureka Server 的互動，用戶端同時具備一個內建的、使用輪詢（round-robin）負載演算法的負載平衡器。在應用啟動後，應用將向 Eureka Server 發送心跳，預設週期為 30 秒，如果 Eureka Server 在多個心跳週期內沒有接收到某個節點的心跳，Eureka Server 會從服務註冊清單中把這個服務節點移除（預設 90 秒）。Eureka Client 分為兩個角色：Application Service（Service Provider）和 Application Client（Service Consumer）。服務提供方是註冊到 Eureka Server 中的服務，服務消費方是透過 Eureka Server 發現服務，並消費。Eureka Server 的架構如圖 5.1 所示。

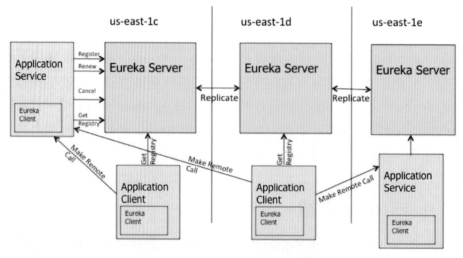

圖 5.1 Eureka Server 的架構圖

- Register（服務註冊）：把自己的 IP 位址和通訊埠註冊給 Eureka。
- Renew（服務續約）：發送心跳封包，每 30 秒發送一次，告訴 Eureka 自己還活著。
- Cancel（服務下線）：當 Provider 關閉時會向 Eureka 發送訊息，把自己從服務清單中刪除，防止 Consumer 呼叫到不存在的服務。
- Get Registry（取得服務註冊清單）：取得其他服務清單。
- Replicate（叢集中資料同步）：Eureka 叢集中的資料複製與同步。
- Make Remote Call（遠端呼叫）：完成服務的遠端呼叫。

5.1.2 Kotlin 整合 Eureka 服務註冊

新增一個 Maven 子專案 chapter05-eureka，這是一個 Eureka Server，其 pom.xml 檔案主要定義了 Eureka Server 相關的套件：

```xml
1.  <?xml version="1.0" encoding="UTF-8"?>
2.  <project xmlns="http://maven.apache.org/POM/4.0.0"
3.          xmlns:xsi="http://www.w3.org/2001/XMLSchema-instance"
4.          xsi:schemaLocation="http://maven.apache.org/POM/4.0.0 http://
    maven.apache.org/xsd/maven-4.0.0.xsd">
5.      <!-- 父 pom -->
6.      <parent>
7.          <artifactId>kotlinspringboot</artifactId>
8.          <groupId>io.kang.kotlinspringboot</groupId>
9.          <version>0.0.1-SNAPSHOT</version>
10.     </parent>
11.     <modelVersion>4.0.0</modelVersion>
12.     <!-- 子專案名稱 -->
13.     <artifactId>chapter05-eureka</artifactId>
14.
15.     <dependencies>
16.         <!-- Eureka 服務端相依套件 -->
17.         <dependency>
```

```
18.              <groupId>org.springframework.cloud</groupId>
19.              <artifactId>spring-cloud-starter-netflix-eureka-server
    </artifactId>
20.              <version>2.2.1.RELEASE</version>
21.          </dependency>
22.          <!-- Kotlin 相關相依套件 -->
23.          <dependency>
24.              <groupId>com.fasterxml.jackson.module</groupId>
25.              <artifactId>jackson-module-kotlin</artifactId>
26.          </dependency>
27.          <dependency>
28.              <groupId>org.jetbrains.kotlin</groupId>
29.              <artifactId>kotlin-reflect</artifactId>
30.          </dependency>
31.          <dependency>
32.              <groupId>org.jetbrains.kotlin</groupId>
33.              <artifactId>kotlin-stdlib-jdk8</artifactId>
34.          </dependency>
35.          <dependency>
36.              <groupId>org.jetbrains.kotlinx</groupId>
37.              <artifactId>kotlinx-coroutines-core</artifactId>
38.              <version>1.3.2</version>
39.          </dependency>
40.      </dependencies>
41.
42.      <build>
43.          <sourceDirectory>${project.basedir}/src/main/kotlin
    </sourceDirectory>
44.          <testSourceDirectory>${project.basedir}/src/test/kotlin
    </testSourceDirectory>
45.          <plugins>
46.              <plugin>
47.                  <groupId>org.springframework.boot</groupId>
48.                  <artifactId>spring-boot-maven-plugin</artifactId>
49.              </plugin>
```

```
50.          <plugin>
51.              <groupId>org.jetbrains.kotlin</groupId>
52.              <artifactId>kotlin-maven-plugin</artifactId>
53.              <configuration>
54.                  <args>
55.                      <arg>-Xjsr305=strict</arg>
56.                  </args>
57.                  <compilerPlugins>
58.                      <plugin>spring</plugin>
59.                      <plugin>jpa</plugin>
60.                  </compilerPlugins>
61.              </configuration>
62.              <dependencies>
63.                  <dependency>
64.                      <groupId>org.jetbrains.kotlin</groupId>
65.                      <artifactId>kotlin-maven-allopen</artifactId>
66.                      <version>${kotlin.version}</version>
67.                  </dependency>
68.                  <dependency>
69.                      <groupId>org.jetbrains.kotlin</groupId>
70.                      <artifactId>kotlin-maven-noarg</artifactId>
71.                      <version>${kotlin.version}</version>
72.                  </dependency>
73.              </dependencies>
74.          </plugin>
75.      </plugins>
76.  </build>
77. </project>
```

application.yml 的定義如下：

```
1.  server:
2.    port: 8761   # 應用通訊埠編號
3.  eureka:
4.    instance:
5.      hostname: localhost   # Eureka 實例的主機
```

```
6.    client:
7.      register-with-eureka: false  # 應用不註冊到 Eureka
8.      fetch-registry: false  # 不從 Eureka 取得其他服務的位址
9.      service-url:  # Eureka 服務位址
10.        defaultZone: http://${eureka.instance.hostname}:${server.port}/
    eureka/
```

EurekaServerApplication.kt 定義了啟動類別：

```
1.  // 開啟 Eureka Server 註釋
2.  @EnableEurekaServer
3.  @SpringBootApplication
4.  class EurekaServerApplication
5.  // 啟動類別
6.  fun main(args: Array<String>) {
7.      runApplication<EurekaServerApplication>(*args);
8.  }
```

執行這個類別，啟動一個單機的 Eureka Server。

5.1.3 一個 Eureka 服務提供方

新增一個 Maven 子專案：chapter05-eureka-provider，這是一個 Eureka Client，定義一個服務提供方。pom.xml 檔案內容如下：

```
1.  <?xml version="1.0" encoding="UTF-8"?>
2.  <project xmlns="http://maven.apache.org/POM/4.0.0"
3.          xmlns:xsi="http://www.w3.org/2001/XMLSchema-instance"
4.          xsi:schemaLocation="http://maven.apache.org/POM/4.0.0 http://
    maven.apache.org/xsd/maven-4.0.0.xsd">
5.      <!-- 父 pom -->
6.      <parent>
7.          <artifactId>kotlinspringboot</artifactId>
8.          <groupId>io.kang.kotlinspringboot</groupId>
9.          <version>0.0.1-SNAPSHOT</version>
```

```
10.    </parent>
11.    <modelVersion>4.0.0</modelVersion>
12.    <!-- 子專案名稱 -->
13.    <artifactId>chapter05-eureka-provider</artifactId>
14.
15.    <dependencies>
16.        <!-- Eureka 用戶端相依套件 -->
17.        <dependency>
18.            <groupId>org.springframework.cloud</groupId>
19.            <artifactId>spring-cloud-starter-netflix-eureka-client
   </artifactId>
20.            <version>2.2.1.RELEASE</version>
21.        </dependency>
22.        <!-- Spring Boot Web 相依套件 -->
23.        <dependency>
24.            <groupId>org.springframework.boot</groupId>
25.            <artifactId>spring-boot-starter-web</artifactId>
26.            <version>2.2.1.RELEASE</version>
27.        </dependency>
28.        <dependency>
29.            <groupId>com.fasterxml.jackson.module</groupId>
30.            <artifactId>jackson-module-kotlin</artifactId>
31.        </dependency>
32.        <dependency>
33.            <groupId>org.jetbrains.kotlin</groupId>
34.            <artifactId>kotlin-reflect</artifactId>
35.        </dependency>
36.        <dependency>
37.            <groupId>org.jetbrains.kotlin</groupId>
38.            <artifactId>kotlin-stdlib-jdk8</artifactId>
39.        </dependency>
40.        <dependency>
41.            <groupId>org.jetbrains.kotlinx</groupId>
42.            <artifactId>kotlinx-coroutines-core</artifactId>
43.            <version>1.3.2</version>
```

```
44.          </dependency>
45.     </dependencies>
46.
47.     <build>
48.          <sourceDirectory>${project.basedir}/src/main/kotlin
    </sourceDirectory>
49.          <testSourceDirectory>${project.basedir}/src/test/kotlin
    </testSourceDirectory>
50.          <plugins>
51.              <plugin>
52.                  <groupId>org.springframework.boot</groupId>
53.                  <artifactId>spring-boot-maven-plugin</artifactId>
54.              </plugin>
55.              <plugin>
56.                  <groupId>org.jetbrains.kotlin</groupId>
57.                  <artifactId>kotlin-maven-plugin</artifactId>
58.                  <configuration>
59.                      <args>
60.                          <arg>-Xjsr305=strict</arg>
61.                      </args>
62.                      <compilerPlugins>
63.                          <plugin>spring</plugin>
64.                          <plugin>jpa</plugin>
65.                      </compilerPlugins>
66.                  </configuration>
67.                  <dependencies>
68.                      <dependency>
69.                          <groupId>org.jetbrains.kotlin</groupId>
70.                          <artifactId>kotlin-maven-allopen</artifactId>
71.                          <version>${kotlin.version}</version>
72.                      </dependency>
73.                      <dependency>
74.                          <groupId>org.jetbrains.kotlin</groupId>
75.                          <artifactId>kotlin-maven-noarg</artifactId>
76.                          <version>${kotlin.version}</version>
```

```
77.                    </dependency>
78.                  </dependencies>
79.                </plugin>
80.              </plugins>
81.          </build>
82.  </project>
```

application.yml 的定義如下：

```
1.  server:
2.    port: 8000                # 應用通訊埠編號
3.  spring:
4.    application:
5.      name: provider-server   # 應用名稱
6.  eureka:
7.    client:
8.      service-url:            # Eureka Server 造訪網址
9.        defaultZone: http://localhost:8761/eureka/
```

ProviderApplication.kt 定義了啟動類別，執行它可以啟動一個 Eureka Client，並註冊到 Eureka Server：

```
1.  // 開啟 Eureka Client 註釋，註冊到 Eureka
2.  @EnableEurekaClient
3.  @SpringBootApplication
4.  class ProviderApplication
5.  // 啟動類別
6.  fun main(args: Array<String>) {
7.      runApplication<ProviderApplication>(*args)
8.  }
```

ProviderController.kt 定義了一個介面，提供服務：

```
1.  @RestController
2.  class ProviderController {
3.      // 服務方測試介面
```

```
4.      @GetMapping("/provide")
5.      fun provide(): String {
6.          return "Hello From Provider"
7.      }
8.  }
```

5.1.4 Kotlin 整合 OpenFeign 服務呼叫

新增一個 Maven 子專案：chapter05-eureka-consumer，這是一個 Eureka Client，是一個服務消費者，呼叫 5.1.3 節定義的服務提供方。pom.xml 的內容如下：

```
1.  <?xml version="1.0" encoding="UTF-8"?>
2.  <project xmlns="http://maven.apache.org/POM/4.0.0"
3.          xmlns:xsi="http://www.w3.org/2001/XMLSchema-instance"
4.          xsi:schemaLocation="http://maven.apache.org/POM/4.0.0 http://
    maven.apache.org/xsd/maven-4.0.0.xsd">
5.      <!-- 父 pom -->
6.      <parent>
7.          <artifactId>kotlinspringboot</artifactId>
8.          <groupId>io.kang.kotlinspringboot</groupId>
9.          <version>0.0.1-SNAPSHOT</version>
10.     </parent>
11.     <modelVersion>4.0.0</modelVersion>
12.     <!-- 子專案名稱 -->
13.     <artifactId>chapter05-eureka-consumer</artifactId>
14.     <dependencies>
15.         <!-- Eureka Client 相依套件 -->
16.         <dependency>
17.             <groupId>org.springframework.cloud</groupId>
18.             <artifactId>spring-cloud-starter-netflix-eureka-client
    </artifactId>
19.             <version>2.2.1.RELEASE</version>
20.         </dependency>
```

```
21.          <!-- Spring Boot Web 相依套件 -->
22.      <dependency>
23.          <groupId>org.springframework.boot</groupId>
24.          <artifactId>spring-boot-starter-web</artifactId>
25.          <version>2.2.1.RELEASE</version>
26.      </dependency>
27.          <!--Spring Cloud OpenFeign 相依套件 -->
28.      <dependency>
29.          <groupId>org.springframework.cloud</groupId>
30.          <artifactId>spring-cloud-starter-openfeign</artifactId>
31.          <version>2.2.1.RELEASE</version>
32.      </dependency>
33.      <dependency>
34.          <groupId>com.fasterxml.jackson.module</groupId>
35.          <artifactId>jackson-module-kotlin</artifactId>
36.      </dependency>
37.      <dependency>
38.          <groupId>org.jetbrains.kotlin</groupId>
39.          <artifactId>kotlin-reflect</artifactId>
40.      </dependency>
41.      <dependency>
42.          <groupId>org.jetbrains.kotlin</groupId>
43.          <artifactId>kotlin-stdlib-jdk8</artifactId>
44.      </dependency>
45.      <dependency>
46.          <groupId>org.jetbrains.kotlinx</groupId>
47.          <artifactId>kotlinx-coroutines-core</artifactId>
48.          <version>1.3.2</version>
49.      </dependency>
50.  </dependencies>
51.
52.  <build>
53.      <sourceDirectory>${project.basedir}/src/main/kotlin
    </sourceDirectory>
54.      <testSourceDirectory>${project.basedir}/src/test/kotlin
```

```
        </testSourceDirectory>
55.         <plugins>
56.             <plugin>
57.                 <groupId>org.springframework.boot</groupId>
58.                 <artifactId>spring-boot-maven-plugin</artifactId>
59.             </plugin>
60.             <plugin>
61.                 <groupId>org.jetbrains.kotlin</groupId>
62.                 <artifactId>kotlin-maven-plugin</artifactId>
63.                 <configuration>
64.                     <args>
65.                         <arg>-Xjsr305=strict</arg>
66.                     </args>
67.                     <compilerPlugins>
68.                         <plugin>spring</plugin>
69.                         <plugin>jpa</plugin>
70.                     </compilerPlugins>
71.                 </configuration>
72.                 <dependencies>
73.                     <dependency>
74.                         <groupId>org.jetbrains.kotlin</groupId>
75.                         <artifactId>kotlin-maven-allopen</artifactId>
76.                         <version>${kotlin.version}</version>
77.                     </dependency>
78.                     <dependency>
79.                         <groupId>org.jetbrains.kotlin</groupId>
80.                         <artifactId>kotlin-maven-noarg</artifactId>
81.                         <version>${kotlin.version}</version>
82.                     </dependency>
83.                 </dependencies>
84.             </plugin>
85.         </plugins>
86.     </build>
87. </project>
```

application.yml 的內容如下：

```
1.  server:
2.    port: 8001              # 應用通訊埠編號
3.  spring:
4.    application:
5.      name: consumer-feign  # 應用名稱
6.  eureka:
7.    client:
8.      service-url:          # Eureka 服務造訪網址
9.        defaultZone: http://localhost:8761/eureka/
```

ConsumerApplication.kt 定義了一個啟動類別，啟動了一個消費者，採用 Feign 進行服務間呼叫，並註冊到 Eureka Server：

```
1.  // 開啟 Feign 和 Eureka Client 註釋
2.  @EnableFeignClients
3.  @EnableEurekaClient
4.  @SpringBootApplication
5.  class ConsumerApplication
6.  // 啟動類別
7.  fun main(args: Array<String>) {
8.      runApplication<ConsumerApplication>(*args)
9.  }
```

ProviderService.kt 定義了一個介面，呼叫 chapter05-eureka-provider 定義的介面，@FeignClient 指定要呼叫的服務名稱：

```
1.  // 透過 Feign 呼叫 provider-server 服務
2.  @FeignClient(value = "provider-server")
3.  interface ProviderService {
4.      @GetMapping("/provide")
5.      fun provide(): String
6.  }
```

ConsumerController.kt 定義了一個介面，可以測試 Feign 呼叫：

```kotlin
1.  @RestController
2.  class ConsumerController {
3.      @Autowired
4.      lateinit var providerService: ProviderService
5.      // 測試介面
6.      @GetMapping("/feignProvide")
7.      fun openProvide(): String {
8.          return providerService.provide()
9.      }
10. }
```

5.1.5 Kotlin 整合 Ribbon 服務呼叫

新增一個 Maven 子專案：chapter05-eureka-consumer-ribbon，它是一個服務消費方，採用 Ribbon 方式消費 5.1.3 節定義的服務。pom 檔案如下：

```xml
1.  <?xml version="1.0" encoding="UTF-8"?>
2.  <project xmlns="http://maven.apache.org/POM/4.0.0"
3.          xmlns:xsi="http://www.w3.org/2001/XMLSchema-instance"
4.          xsi:schemaLocation="http://maven.apache.org/POM/4.0.0 http://
    maven.apache.org/xsd/maven-4.0.0.xsd">
5.      <!-- 父 pom -->
6.      <parent>
7.          <artifactId>kotlinspringboot</artifactId>
8.          <groupId>io.kang.kotlinspringboot</groupId>
9.          <version>0.0.1-SNAPSHOT</version>
10.     </parent>
11.     <modelVersion>4.0.0</modelVersion>
12.     <!-- 子專案名 -->
13.     <artifactId>chapter05-eureka-consumer-ribbon</artifactId>
14.     <dependencies>
15.         <!-- Eureka Client 相依套件 -->
16.         <dependency>
```

```xml
17.             <groupId>org.springframework.cloud</groupId>
18.             <artifactId>spring-cloud-starter-netflix-eureka-client
   </artifactId>
19.             <version>2.2.1.RELEASE</version>
20.         </dependency>
21.         <!-- Spring Boot Web 相依套件 -->
22.     <dependency>
23.             <groupId>org.springframework.boot</groupId>
24.             <artifactId>spring-boot-starter-web</artifactId>
25.             <version>2.2.1.RELEASE</version>
26.         </dependency>
27.         <!-- Spring Cloud Ribbon 相依套件 -->
28.     <dependency>
29.             <groupId>org.springframework.cloud</groupId>
30.             <artifactId>spring-cloud-starter-netflix-ribbon</artifactId>
31.             <version>2.2.1.RELEASE</version>
32.         </dependency>
33.     <dependency>
34.             <groupId>com.fasterxml.jackson.module</groupId>
35.             <artifactId>jackson-module-kotlin</artifactId>
36.         </dependency>
37.     <dependency>
38.             <groupId>org.jetbrains.kotlin</groupId>
39.             <artifactId>kotlin-reflect</artifactId>
40.         </dependency>
41.     <dependency>
42.             <groupId>org.jetbrains.kotlin</groupId>
43.             <artifactId>kotlin-stdlib-jdk8</artifactId>
44.         </dependency>
45.     <dependency>
46.             <groupId>org.jetbrains.kotlinx</groupId>
47.             <artifactId>kotlinx-coroutines-core</artifactId>
48.             <version>1.3.2</version>
49.         </dependency>
50.     </dependencies>
```

```
51.    <build>
52.        <sourceDirectory>${project.basedir}/src/main/kotlin
   </sourceDirectory>
53.        <testSourceDirectory>${project.basedir}/src/test/kotlin
   </testSourceDirectory>
54.        <plugins>
55.            <plugin>
56.                <groupId>org.springframework.boot</groupId>
57.                <artifactId>spring-boot-maven-plugin</artifactId>
58.            </plugin>
59.            <plugin>
60.                <groupId>org.jetbrains.kotlin</groupId>
61.                <artifactId>kotlin-maven-plugin</artifactId>
62.                <configuration>
63.                    <args>
64.                        <arg>-Xjsr305=strict</arg>
65.                    </args>
66.                    <compilerPlugins>
67.                        <plugin>spring</plugin>
68.                        <plugin>jpa</plugin>
69.                    </compilerPlugins>
70.                </configuration>
71.                <dependencies>
72.                    <dependency>
73.                        <groupId>org.jetbrains.kotlin</groupId>
74.                        <artifactId>kotlin-maven-allopen</artifactId>
75.                        <version>${kotlin.version}</version>
76.                    </dependency>
77.                    <dependency>
78.                        <groupId>org.jetbrains.kotlin</groupId>
79.                        <artifactId>kotlin-maven-noarg</artifactId>
80.                        <version>${kotlin.version}</version>
81.                    </dependency>
82.                </dependencies>
83.            </plugin>
```

```
84.        </plugins>
85.     </build>
86. </project>
```

application.yml 的內容如下：

```
1.  server:
2.    port: 8002              # 應用通訊埠編號
3.  spring:
4.    application:
5.      name: consumer-ribbon     # 應用名稱
6.  eureka:
7.    client:
8.      service-url:               # Eureka Server 造訪網址
9.        defaultZone: http://localhost:8761/eureka/
```

RibbonApplication.kt 定義了啟動類別，定義一個 restTemplate 和輪詢策略，當它啟動後會註冊到 Eureka Server：

```
1.  // 開啟 Eureka Client 註釋
2.  @SpringBootApplication
3.  @EnableEurekaClient
4.  class RibbonApplication {
5.      // 建立 restTemplate，進行負載平衡
6.      @Bean
7.      @LoadBalanced
8.      fun restTemplate(): RestTemplate {
9.          return RestTemplate()
10.     }
11.     // 負載平衡採用隨機分配規則
12.     @Bean
13.     fun ribbonRule(): IRule {
14.         return RandomRule()
15.     }
16. }
17. // 啟動類別
```

```
18. fun main(args: Array<String>) {
19.     runApplication<RibbonApplication>(*args)
20. }
```

RibbonService.kt 定義了一個方法，採用 Ribbon 方式存取 chapter05-eureka-provider 定義的介面：

```
1.  @Component
2.  class RibbonService {
3.
4.      @Autowired
5.      lateinit var restTemplate: RestTemplate
6.      // 採用 Ribbon 方式呼叫 provider-server 的 provide 介面
7.      fun ribbonProvide(): String? {
8.          return restTemplate.getForObject("http://PROVIDER-SERVER/provide",
    String::class.java)
9.      }
10. }
```

RibbonController.kt 定義了一個介面，用於測試存取服務提供者的介面：

```
1.  @RestController
2.  class RibbonController {
3.      @Autowired
4.      lateinit var ribbonService: RibbonService
5.      // 測試介面
6.      @GetMapping("/ribbonProvide")
7.      fun ribbonProvide(): String? {
8.          return ribbonService.ribbonProvide()
9.      }
10. }
```

依次啟動 chapter05-eureka、chapter05-eureka-provider、chapter05-eureka-consumer、chapter05-eureka-consumer-ribbon，可以看到如圖 5.2 所示的服務清單。

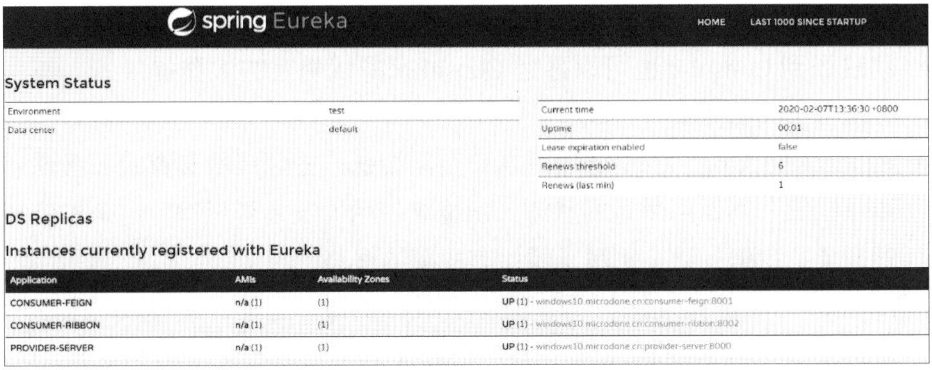

圖 5.2　Eureka Server 註冊的服務清單

一個服務提供方和兩個服務消費方都註冊到了 Eureka Server。呼叫 consumer-feign 的介面 "/feignProvide" 和 consumer-ribbon 的介面 "/ribbonProvide"，都能存取到 provider-server 的介面 "/provide"。

5.2　Consul

Consul 是 HashiCorp 公司推出的開放原始碼產品，用於實現分散式系統的服務發現、服務隔離、服務設定。Consul 內建了服務註冊與發現架構、分佈一致性協定實現、健康檢查、Key-Value 儲存、多資料中心方案，不再需要依賴其他工具。Consul 本身使用 Go 語言開發，具有跨平台、執行高效等特點。本節介紹使用 Consul 作為 Kotlin 開發的微服務的註冊中心的相關知識。

5.2.1　Consul 介紹

Consul 的主要特點有以下幾點。

- Service Discovery：服務註冊與發現，Consul 的用戶端可以作為一個服務註冊到 Consul，也可以透過 Consul 來尋找特定的服務提供者，並且根據提供的資訊進行呼叫。
- Health Checking：Consul 用戶端會定期發送一些健康檢查資料和服務端進行通訊，判斷用戶端的狀態、記憶體使用情況是否正常，用來監控整個叢集的狀態，防止將請求轉給有故障的服務。
- KV Store：Consul 還提供了一個容易使用的鍵值儲存。這可以用來保持動態設定、協助服務協調、建立 Leader 選舉，以及開發者想建置的其他一些交易。
- Secure Service Communication：Consul 可以為服務產生分散式的 TLS 憑證，以建立相互的 TLS 連接。可以使用 intentions 定義允許哪些服務進行通訊，還可以使用 intentions 輕鬆管理服務隔離，而非使用複雜的網路拓撲和靜態防火牆規則。
- Multi Datacenter：Consul 支援開箱即用的多資料中心，這表示使用者不用擔心需要建立額外的抽象層讓業務擴充到多個區域。

Consul 有服務端和用戶端兩個角色，服務端儲存設定資訊，組成高可用叢集，在區域網內與本機用戶端通訊，透過廣域網路與其他資料中心通訊。每個資料中心的伺服器數量推薦為 3 個或是 5 個。用戶端是無狀態的，負責將 HTTP 和 DNS 介面請求轉發給區域網內的服務端叢集。

Consul 的呼叫過程如圖 5.3 所示。

當 Producer 啟動的時候，會向 Consul 發送一個 POST 請求，告訴 Consul 自己的 IP 位址、通訊埠；Consul 接收到 Producer 的註冊後，每隔 10 秒（預設）向 Producer 發送一個健康檢查的請求，檢驗 Producer 是否健康；當 Consumer 發送 GET 方式的請求 /api/address 到 Producer 時，會先從 Consul 中拿到一個儲存服務 IP 位址、通訊埠的臨時表，從表中拿到

Producer 的 IP 位址和通訊埠後再發送 GET 方式的請求 /api/address；該臨時表每隔 10 秒更新一次，只包含透過健康檢查的 Producer。

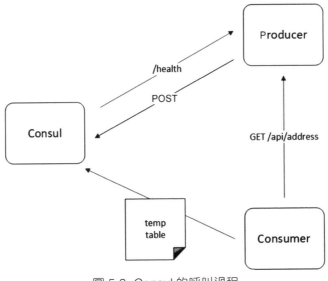

圖 5.3　Consul 的呼叫過程

5.2.2　Kotlin 整合 Consul 服務註冊

在 Consul 官網下載 Consul（Consul_1.6.3_windows_amd64），解壓後執行 Consul agent –dev 以啟動 Consul。

新增一個 Maven 子專案 chapter05-Consul，它是一個服務提供者，也是一個 Consul 用戶端。pom 檔案如下：

```
1.  <?xml version="1.0" encoding="UTF-8"?>
2.  <project xmlns="http://maven.apache.org/POM/4.0.0"
3.         xmlns:xsi="http://www.w3.org/2001/XMLSchema-instance"
4.         xsi:schemaLocation="http://maven.apache.org/POM/4.0.0 http://
    maven.apache.org/xsd/maven-4.0.0.xsd">
5.      <parent>
```

```
6.          <artifactId>kotlinspringboot</artifactId>
7.          <groupId>io.kang.kotlinspringboot</groupId>
8.          <version>0.0.1-SNAPSHOT</version>
9.      </parent>
10.     <modelVersion>4.0.0</modelVersion>
11.     <!-- 子專案名 -->
12.     <artifactId>chapter05-consul</artifactId>
13.
14.     <dependencies>
15.         <!-- Spring Cloud Consul 相依套件 -->
16.         <dependency>
17.             <groupId>org.springframework.cloud</groupId>
18.             <artifactId>spring-cloud-starter-consul-discovery</artifactId>
19.             <version>2.2.1.RELEASE</version>
20.         </dependency>
21.         <-- Spring Boot Web 相依套件 -->
22.         <dependency>
23.             <groupId>org.springframework.boot</groupId>
24.             <artifactId>spring-boot-starter-web</artifactId>
25.             <version>2.2.1.RELEASE</version>
26.         </dependency>
27.         <!-- Spring Boot Actuator 相依套件 -->
28.         <dependency>
29.             <groupId>org.springframework.boot</groupId>
30.             <artifactId>spring-boot-starter-actuator</artifactId>
31.             <version>2.2.1.RELEASE</version>
32.         </dependency>
33.         <dependency>
34.             <groupId>com.fasterxml.jackson.module</groupId>
35.             <artifactId>jackson-module-kotlin</artifactId>
36.         </dependency>
37.         <dependency>
38.             <groupId>org.jetbrains.kotlin</groupId>
39.             <artifactId>kotlin-reflect</artifactId>
40.         </dependency>
```

```xml
41.     <dependency>
42.         <groupId>org.jetbrains.kotlin</groupId>
43.         <artifactId>kotlin-stdlib-jdk8</artifactId>
44.     </dependency>
45.     <dependency>
46.         <groupId>org.jetbrains.kotlinx</groupId>
47.         <artifactId>kotlinx-coroutines-core</artifactId>
48.         <version>1.3.2</version>
49.     </dependency>
50.   </dependencies>
51.   <build>
52.       <sourceDirectory>${project.basedir}/src/main/kotlin</sourceDirectory>
53.       <testSourceDirectory>${project.basedir}/src/test/kotlin</testSourceDirectory>
54.       <plugins>
55.           <plugin>
56.               <groupId>org.springframework.boot</groupId>
57.               <artifactId>spring-boot-maven-plugin</artifactId>
58.           </plugin>
59.           <plugin>
60.               <groupId>org.jetbrains.kotlin</groupId>
61.               <artifactId>kotlin-maven-plugin</artifactId>
62.               <configuration>
63.                   <args>
64.                       <arg>-Xjsr305=strict</arg>
65.                   </args>
66.                   <compilerPlugins>
67.                       <plugin>spring</plugin>
68.                       <plugin>jpa</plugin>
69.                   </compilerPlugins>
70.               </configuration>
71.               <dependencies>
72.                   <dependency>
73.                       <groupId>org.jetbrains.kotlin</groupId>
```

```
74.                    <artifactId>kotlin-maven-allopen</artifactId>
75.                    <version>${kotlin.version}</version>
76.                </dependency>
77.                <dependency>
78.                    <groupId>org.jetbrains.kotlin</groupId>
79.                    <artifactId>kotlin-maven-noarg</artifactId>
80.                    <version>${kotlin.version}</version>
81.                </dependency>
82.            </dependencies>
83.          </plugin>
84.        </plugins>
85.      </build>
86. </project>
```

application.yml 檔案的內容如下：

```
1.  server:
2.    port: 8080
3.  spring:
4.    cloud:
5.      consul:
6.        host: localhost   # consul 主機
7.        port: 8500        # consul 通訊埠
8.        discovery:        # 註冊到 Consul 的服務名稱
9.          service-name: ${spring.application.name}
10.   application:
11.     name: consul-producer
```

ConsulApplication.kt 定義了一個啟動類別，啟動後會註冊到 Consul：

```
1.  // 開啟註釋，服務註冊到 Consul
2.  @SpringBootApplication
3.  @EnableDiscoveryClient
4.  class ConsulApplication
5.  // 啟動函數
6.  fun main(args: Array<String>) {
```

```
7.     runApplication<ConsulApplication>(*args)
8. }
```

ConsulController.kt 定義了一個服務介面，用於測試：

```
1. @RestController
2. class ConsulController {
3.     // 測試介面
4.     @GetMapping("hello/consul")
5.     fun helloConsul(): String {
6.         return "Hello Consul"
7.     }
8. }
```

Consul 的監控介面如圖 5.4 所示。

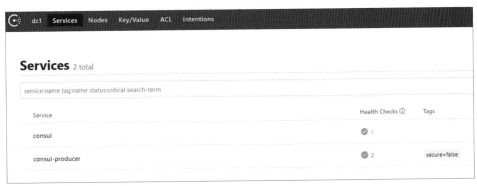

圖 5.4　Consul 的監控介面

5.2.3 Kotlin 整合 OpenFeign 和 Ribbon 服務呼叫

新增一個 Maven 專案：chapter05-consul-consumer，它是一個消費者，採用 Feign、Ribbon 方式呼叫 5.2.2 節定義的服務提供者。pom 檔案的內容如下：

```
1. <?xml version="1.0" encoding="UTF-8"?>
2. <project xmlns="http://maven.apache.org/POM/4.0.0"
3.         xmlns:xsi="http://www.w3.org/2001/XMLSchema-instance"
```

```
4.          xsi:schemaLocation="http://maven.apache.org/POM/4.0.0 http://
   maven.apache.org/xsd/maven-4.0.0.xsd">
5.      <parent>
6.          <artifactId>kotlinspringboot</artifactId>
7.          <groupId>io.kang.kotlinspringboot</groupId>
8.          <version>0.0.1-SNAPSHOT</version>
9.      </parent>
10.     <modelVersion>4.0.0</modelVersion>
11.     <!-- 子專案名 -->
12.     <artifactId>chapter05-consul-consumer</artifactId>
13.     <dependencies>
14.         <!-- Spring Cloud Consul 相依套件 -->
15.         <dependency>
16.             <groupId>org.springframework.cloud</groupId>
17.             <artifactId>spring-cloud-starter-consul-discovery</artifactId>
18.             <version>2.2.1.RELEASE</version>
19.         </dependency>
20.         <!-- Spring Cloud OpenFeign 相依套件 -->
21.         <dependency>
22.             <groupId>org.springframework.cloud</groupId>
23.             <artifactId>spring-cloud-starter-openfeign</artifactId>
24.             <version>2.2.1.RELEASE</version>
25.         </dependency>
26.         <!-- Spring Boot Web 相依套件 -->
27.         <dependency>
28.             <groupId>org.springframework.boot</groupId>
29.             <artifactId>spring-boot-starter-web</artifactId>
30.             <version>2.2.1.RELEASE</version>
31.         </dependency>
32.         <!-- Spring Boot Actuator 相依套件 -->
33.         <dependency>
34.             <groupId>org.springframework.boot</groupId>
35.             <artifactId>spring-boot-starter-actuator</artifactId>
36.             <version>2.2.1.RELEASE</version>
37.         </dependency>
```

```
38.        <dependency>
39.            <groupId>com.fasterxml.jackson.module</groupId>
40.            <artifactId>jackson-module-kotlin</artifactId>
41.        </dependency>
42.        <dependency>
43.            <groupId>org.jetbrains.kotlin</groupId>
44.            <artifactId>kotlin-reflect</artifactId>
45.        </dependency>
46.        <dependency>
47.            <groupId>org.jetbrains.kotlin</groupId>
48.            <artifactId>kotlin-stdlib-jdk8</artifactId>
49.        </dependency>
50.        <dependency>
51.            <groupId>org.jetbrains.kotlinx</groupId>
52.            <artifactId>kotlinx-coroutines-core</artifactId>
53.            <version>1.3.2</version>
54.        </dependency>
55.    </dependencies>
56.    <build>
57.        <sourceDirectory>${project.basedir}/src/main/kotlin
    </sourceDirectory>
58.        <testSourceDirectory>${project.basedir}/src/test/kotlin
    </testSourceDirectory>
59.        <plugins>
60.            <plugin>
61.                <groupId>org.springframework.boot</groupId>
62.                <artifactId>spring-boot-maven-plugin</artifactId>
63.            </plugin>
64.            <plugin>
65.                <groupId>org.jetbrains.kotlin</groupId>
66.                <artifactId>kotlin-maven-plugin</artifactId>
67.                <configuration>
68.                    <args>
69.                        <arg>-Xjsr305=strict</arg>
70.                    </args>
```

```
71.                    <compilerPlugins>
72.                        <plugin>spring</plugin>
73.                        <plugin>jpa</plugin>
74.                    </compilerPlugins>
75.                </configuration>
76.                <dependencies>
77.                    <dependency>
78.                        <groupId>org.jetbrains.kotlin</groupId>
79.                        <artifactId>kotlin-maven-allopen</artifactId>
80.                        <version>${kotlin.version}</version>
81.                    </dependency>
82.                    <dependency>
83.                        <groupId>org.jetbrains.kotlin</groupId>
84.                        <artifactId>kotlin-maven-noarg</artifactId>
85.                        <version>${kotlin.version}</version>
86.                    </dependency>
87.                </dependencies>
88.            </plugin>
89.        </plugins>
90.    </build>
91. </project>
```

application.yml 檔案的內容如下所示：

```
1.  spring:
2.    application:
3.      name: consul-consumer      # 應用名稱
4.    cloud:
5.      consul:
6.        host: 127.0.0.1          # Consul 主機
7.        port: 8500               # Consul 通訊埠
8.        discovery:
9.          register: false        # 僅作為消費者，不註冊服務
10. server:
11.   port: 8005                   # 應用通訊埠編號
```

ConsulConsumer.kt 定義了一個啟動類別，啟動後執行一個 Consul 消費者：

```kotlin
1.  // 開啟 Feign 註釋，開啟服務註冊註釋
2.  @SpringBootApplication
3.  @EnableDiscoveryClient
4.  @EnableFeignClients
5.  class ConsulConsumer {
6.      @Bean
7.      @LoadBalanced
8.      fun restTemplate(): RestTemplate {
9.          return RestTemplate()
10.     }
11. }
12. // 啟動函數
13. fun main(args: Array<String>) {
14.     runApplication<ConsulConsumer>(*args)
15. }
```

FeignService.kt 定義了一個 Feign 介面，用於呼叫 "hello/consul" 這個介面：

```kotlin
1.  @FeignClient(value = "consul-producer")
2.  interface FeignService {
3.      // 測試介面，透過 Feign 方式呼叫
4.      @GetMapping("hello/consul")
5.      fun helloConsul(): String
6.  }
```

ConsumerController.kt 定義了 Feign 和 Ribbon 的測試介面：

```kotlin
1.  @RestController
2.  class ConsumerController {
3.      @Autowired
4.      lateinit var restTemplate: RestTemplate
5.      @Autowired
6.      lateinit var feignService: FeignService
7.      // 測試介面，透過 Ribbon 方式呼叫
8.      @GetMapping("ribbon/hello/consul")
```

```
9.      fun ribbonHelloConsul(): String? {
10.         return restTemplate.getForObject("http://consul-producer/hello/
    consul", String::class.java)
11.     }
12.     // 測試介面，透過 Feign 方式呼叫
13.     @GetMapping("feign/hello/consul")
14.     fun feignHelloConsul(): String {
15.         return feignService.helloConsul()
16.     }
17. }
```

服務啟動後，透過 "ribbon/hello/consul" 採用 Ribbon 方式可以存取 "hello/
consul" 這個服務介面。透過 "feign/hello/consul" 採用 Feign 方式也可以存
取 "hello/consul" 這個服務介面。

5.3 Zookeeper

Zookeeper 是一個分散式的、開放原始碼的程式協調服務，是 Hadoop 專案
下的子專案。它提供的主要功能包含：設定管理、名稱服務、分散式鎖、
叢集管理。本節介紹使用 Zookeeper 作為 Kotlin 開發的微服務的註冊中心
的相關知識。

5.3.1 Zookeeper 介紹

Zookeeper 使用 Zab 協定來提供一致性。很多開放原始碼專案使用
Zookeeper 來維護設定，舉例來說，HBase 的用戶端透過 Zookeeper 獲
得必要的 HBase 叢集的設定資訊，開放原始碼訊息佇列 Kafka 使用
Zookeeper 維護 Broker 的資訊，Alibaba 開放原始碼的 SOA 架構 Dubbo 使
用 Zookeeper 管理一些設定來實現服務治理。

Zookeeper 的命名服務功能主要是根據指定名字來取得資源或服務的位址、提供者等資訊，利用其 znode 的特點和 watcher 機制，將其作為動態註冊和取得服務資訊的設定中心，統一管理服務名稱和其對應的伺服器列表資訊，能夠近乎即時地感知到後端伺服器的狀態（上線、下線、當機）。

在分散式環境中，為了加強可用性，叢集中的每台伺服器上都部署著同樣的服務。叢集中的每台伺服器需要協調，使用分散式鎖，在某個時刻只讓一台伺服器工作，當這台伺服器出問題的時候釋放鎖。首先需要建立一個父節點，儘量是持久節點，然後每個要獲得鎖的執行緒都會在這個節點下建立一個臨時順序節點，由於序號的遞增性，可以規定排號最小的那個獲得鎖。Zookeeper 的節點監聽機制可以確保佔有鎖的方式有序而且高效。

Zookeeper 可以管理叢集和服務發現。在分散式的叢集中，經常會由於各種原因，例如硬體故障、軟體故障、網路問題，有些節點會進進出出。有新的節點加入進來，也有老的節點退出叢集。這個時候，叢集中的其他機器需要能感知到這種變化，然後根據這種變化做出對應的決策。當消費者存取某個服務時，就需要採用某種機制發現現在有哪些節點可以提供該服務。

在生產環境中，ZooKeeper 採用叢集部署。Zookeeper 叢集中包含 Leader、Follower 以及 Observer 三個角色。Leader 負責進行投票的發起和決議，更新系統狀態，Leader 由選舉產生；Follower 用於接收用戶端請求並向用戶端傳回結果，在選舉過程中參與投票；Observer 可以接收用戶端連接，接收讀寫入請求，將寫入請求轉發給 Leader，但 Observer 不參加投票過程，只同步 Leader 的狀態，Observer 的目的是為了擴充系統，加強讀取速度。在 Zookeeper 系統中，只要叢集中存在超過一半的節點（這裡指的是投票節點即非 Observer 節點）能夠正常執行，那麼整個叢集就能夠正常對外服務。

Zookeeper 的叢集架構如圖 5.5 所示。

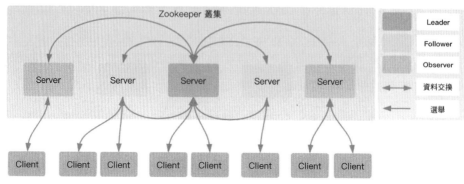

圖 5.5 Zookeeper 的叢集架構

5.3.2 Kotlin 整合 Zookeeper 服務註冊

新增一個子專案:chapter05-zk,這是一個服務提供者,pom 檔案如下:

```
1.  <?xml version="1.0" encoding="UTF-8"?>
2.  <project xmlns="http://maven.apache.org/POM/4.0.0"
3.          xmlns:xsi="http://www.w3.org/2001/XMLSchema-instance"
4.          xsi:schemaLocation="http://maven.apache.org/POM/4.0.0 http://
    maven.apache.org/xsd/maven-4.0.0.xsd">
5.      <parent>
6.          <artifactId>kotlinspringboot</artifactId>
7.          <groupId>io.kang.kotlinspringboot</groupId>
8.          <version>0.0.1-SNAPSHOT</version>
9.      </parent>
10.     <modelVersion>4.0.0</modelVersion>
11.     <!-- 子專案名 -->
12.     <artifactId>chapter05-zk</artifactId>
13.     <dependencies>
14.         <!-- Spring Cloud Zookeeper 相依套件 -->
15.         <dependency>
16.             <groupId>org.springframework.cloud</groupId>
17.             <artifactId>spring-cloud-starter-zookeeper-discovery
    </artifactId>
```

```xml
18.          <version>2.2.0.RELEASE</version>
19.      </dependency>
20.      <!-- Spring Boot Web 相依套件 -->
21.      <dependency>
22.          <groupId>org.springframework.boot</groupId>
23.          <artifactId>spring-boot-starter-web</artifactId>
24.          <version>2.2.1.RELEASE</version>
25.      </dependency>
26.      <dependency>
27.          <groupId>com.fasterxml.jackson.module</groupId>
28.          <artifactId>jackson-module-kotlin</artifactId>
29.      </dependency>
30.      <dependency>
31.          <groupId>org.jetbrains.kotlin</groupId>
32.          <artifactId>kotlin-reflect</artifactId>
33.      </dependency>
34.      <dependency>
35.          <groupId>org.jetbrains.kotlin</groupId>
36.          <artifactId>kotlin-stdlib-jdk8</artifactId>
37.      </dependency>
38.      <dependency>
39.          <groupId>org.jetbrains.kotlinx</groupId>
40.          <artifactId>kotlinx-coroutines-core</artifactId>
41.          <version>1.3.2</version>
42.      </dependency>
43.   </dependencies>
44.   <build>
45.      <sourceDirectory>${project.basedir}/src/main/kotlin</sourceDirectory>
46.      <testSourceDirectory>${project.basedir}/src/test/kotlin</testSourceDirectory>
47.      <plugins>
48.          <plugin>
49.              <groupId>org.springframework.boot</groupId>
50.              <artifactId>spring-boot-maven-plugin</artifactId>
```

```
51.            </plugin>
52.            <plugin>
53.                <groupId>org.jetbrains.kotlin</groupId>
54.                <artifactId>kotlin-maven-plugin</artifactId>
55.                <configuration>
56.                    <args>
57.                        <arg>-Xjsr305=strict</arg>
58.                    </args>
59.                    <compilerPlugins>
60.                        <plugin>spring</plugin>
61.                        <plugin>jpa</plugin>
62.                    </compilerPlugins>
63.                </configuration>
64.                <dependencies>
65.                    <dependency>
66.                        <groupId>org.jetbrains.kotlin</groupId>
67.                        <artifactId>kotlin-maven-allopen</artifactId>
68.                        <version>${kotlin.version}</version>
69.                    </dependency>
70.                    <dependency>
71.                        <groupId>org.jetbrains.kotlin</groupId>
72.                        <artifactId>kotlin-maven-noarg</artifactId>
73.                        <version>${kotlin.version}</version>
74.                    </dependency>
75.                </dependencies>
76.            </plugin>
77.        </plugins>
78.    </build>
79. </project>
```

application.yml 檔案的內容如下所示：

```
1.  server:
2.    port: 8081                  # 應用通訊埠編號
3.  spring:
4.    application:
```

```
5.     name: zk-produce        # 應用名稱
6.   cloud:
7.     zookeeper:              #Zookeeper 的主機、通訊埠
8.       connect-string: localhost:2181
```

ZkApplication.kt 定義了一個啟動類別，服務啟動後，會註冊到 Zookeeper：

```
1.  // 開啟服務註冊註釋
2.  @SpringBootApplication
3.  @EnableDiscoveryClient
4.  class ZkApplication
5.  // 啟動函數
6.  fun main(args: Array<String>) {
7.      runApplication<ZkApplication>(*args)
8.  }
```

服務在 Zookeeper 中的註冊資訊如下所示，包含應用名、id、address、port 等資訊：

```
1.  {
2.      "name": "zk-produce",
3.      "id": "133ff0ac-cd6f-4942-886c-b76243df7550",
4.      "address": "windows10.microdone.cn",
5.      "port": 8081,
6.      "sslPort": null,
7.      "payload": {
8.          "@class": "org.springframework.cloud.zookeeper.discovery.
    ZookeeperInstance",
9.          "id": "application-1",
10.         "name": "zk-produce",
11.         "metadata": {}
12.     },
13.     "registrationTimeUTC": 1581241617714,
14.     "serviceType": "DYNAMIC",
15.     "uriSpec": {
16.         "parts": [{
17.             "value": "scheme",
```

```
18.                 "variable": true
19.             }, {
20.                 "value": "://",
21.                 "variable": false
22.             }, {
23.                 "value": "address",
24.                 "variable": true
25.             }, {
26.                 "value": ":",
27.                 "variable": false
28.             }, {
29.                 "value": "port",
30.                 "variable": true
31.             }]
32.         }
33. }
```

ZkController.kt 定義了一個測試介面：

```
1.  @RestController
2.  class ZkController {
3.      // 測試介面
4.      @GetMapping("hello/zk")
5.      fun helloZk(): String {
6.          return "hello zookeeper"
7.      }
8.  }
```

5.3.3 Kotlin 整合 OpenFeign 和 Ribbon 服務呼叫

新增一個子專案：chapter05-zk-consumer，這是一個服務消費者，pom 檔案如下：

```
1.  <?xml version="1.0" encoding="UTF-8"?>
2.  <project xmlns="http://maven.apache.org/POM/4.0.0"
3.           xmlns:xsi="http://www.w3.org/2001/XMLSchema-instance"
```

```
4.          xsi:schemaLocation="http://maven.apache.org/POM/4.0.0 http://
   maven.apache.org/xsd/maven-4.0.0.xsd">
5.      <parent>
6.          <artifactId>kotlinspringboot</artifactId>
7.          <groupId>io.kang.kotlinspringboot</groupId>
8.          <version>0.0.1-SNAPSHOT</version>
9.      </parent>
10.     <modelVersion>4.0.0</modelVersion>
11.     <!-- 子專案名 -->
12.     <artifactId>chapter05-zk-consumer</artifactId>
13.     <dependencies>
14.         <!-- Spring Cloud Zookeeper 相依套件 -->
15.         <dependency>
16.             <groupId>org.springframework.cloud</groupId>
17.             <artifactId>spring-cloud-starter-zookeeper-discovery
   </artifactId>
18.             <version>2.2.0.RELEASE</version>
19.         </dependency>
20.         <!-- Spring Boot Web 相依套件 -->
21.         <dependency>
22.             <groupId>org.springframework.boot</groupId>
23.             <artifactId>spring-boot-starter-web</artifactId>
24.             <version>2.2.1.RELEASE</version>
25.         </dependency>
26.         <!-- Spring Cloud OpenFeign 相依套件 -->
27.         <dependency>
28.             <groupId>org.springframework.cloud</groupId>
29.             <artifactId>spring-cloud-starter-openfeign</artifactId>
30.             <version>2.2.1.RELEASE</version>
31.         </dependency>
32.         <dependency>
33.             <groupId>com.fasterxml.jackson.module</groupId>
34.             <artifactId>jackson-module-kotlin</artifactId>
35.         </dependency>
36.         <dependency>
```

```
37.              <groupId>org.jetbrains.kotlin</groupId>
38.              <artifactId>kotlin-reflect</artifactId>
39.          </dependency>
40.          <dependency>
41.              <groupId>org.jetbrains.kotlin</groupId>
42.              <artifactId>kotlin-stdlib-jdk8</artifactId>
43.          </dependency>
44.          <dependency>
45.              <groupId>org.jetbrains.kotlinx</groupId>
46.              <artifactId>kotlinx-coroutines-core</artifactId>
47.              <version>1.3.2</version>
48.          </dependency>
49.      </dependencies>
50.      <build>
51.          <sourceDirectory>${project.basedir}/src/main/kotlin</sourceDirectory>
52.          <testSourceDirectory>${project.basedir}/src/test/kotlin
     </testSourceDirectory>
53.          <plugins>
54.              <plugin>
55.                  <groupId>org.springframework.boot</groupId>
56.                  <artifactId>spring-boot-maven-plugin</artifactId>
57.              </plugin>
58.              <plugin>
59.                  <groupId>org.jetbrains.kotlin</groupId>
60.                  <artifactId>kotlin-maven-plugin</artifactId>
61.                  <configuration>
62.                      <args>
63.                          <arg>-Xjsr305=strict</arg>
64.                      </args>
65.                      <compilerPlugins>
66.                          <plugin>spring</plugin>
67.                          <plugin>jpa</plugin>
68.                      </compilerPlugins>
69.                  </configuration>
70.                  <dependencies>
```

```
71.                    <dependency>
72.                        <groupId>org.jetbrains.kotlin</groupId>
73.                        <artifactId>kotlin-maven-allopen</artifactId>
74.                        <version>${kotlin.version}</version>
75.                    </dependency>
76.                    <dependency>
77.                        <groupId>org.jetbrains.kotlin</groupId>
78.                        <artifactId>kotlin-maven-noarg</artifactId>
79.                        <version>${kotlin.version}</version>
80.                    </dependency>
81.                </dependencies>
82.            </plugin>
83.        </plugins>
84.    </build>
85. </project>
```

application.yml 檔案的內容如下：

```
1. server:
2.   port: 8006              # 應用通訊埠編號
3. spring:
4.   application:
5.     name: zk-consumer    # 應用名稱
6.   cloud:
7.     zookeeper:           #Zookeeper 連接位址
8.       connect-string: localhost:2181
```

ZkConsumerApplication.kt 定義了啟動類別，服務啟動後，會註冊到 Zookeeper：

```
1. // 開啟 Feign 註釋，開啟服務註冊註釋
2. @EnableDiscoveryClient
3. @SpringBootApplication
4. @EnableFeignClients
5. class ZkConsumerApplication {
6.     @Bean
7.     @LoadBalanced
```

```
8.      fun restTemplate():RestTemplate {
9.          return RestTemplate()
10.     }
11. }
12. // 啟動函數
13. fun main(args: Array<String>) {
14.     runApplication<ZkConsumerApplication>(*args)
15. }
```

FeignService.kt 定義了一個測試介面，使用 Feign 方式呼叫 zk-produce 的 hello/zk 介面：

```
1.  @FeignClient(value = "zk-produce", path = "/")
2.  @Component
3.  interface FeignService {
4.      // 測試介面，使用 feign 方式
5.      @GetMapping("hello/zk")
6.      fun helloZk(): String
7.  }
```

ZkConsumerController.kt 定義了兩個測試介面，分別用 Ribbon、Feign 方式呼叫 zk-produce 的 hello/zk 介面：

```
1.  @RestController
2.  class ZkConsumerController {
3.      @Autowired
4.      lateinit var feignService: FeignService
5.
6.      @Autowired
7.      lateinit var restTemplate: RestTemplate
8.      // 測試介面，使用 Feign 方式
9.      @GetMapping("feign/hello/zk")
10.     fun feignHelloZk(): String {
11.         return feignService.helloZk()
12.     }
13.     // 測試介面，使用 Ribbon 方式
```

```
14.     @GetMapping("ribbon/hello/zk")
15.     fun ribbonHelloZk(): String? {
16.         return restTemplate.getForObject("http://zk-produce/hello/zk",
   String::class.java)
17.     }
18. }
```

5.4 Nacos

Nacos 是阿里巴巴開放原始碼的服務發現、設定和管理工具。Nacos 提供了一組簡單好用的特性集，有助快速實現動態服務發現、服務設定、服務中繼資料及流量管理。本節介紹使用 Nacos 作為 Kotlin 開發的微服務註冊中心的相關知識。

5.4.1 Nacos 介紹

Nacos 的特性有以下幾點。

服務發現和服務健康監測：Nacos 支援以 DNS 和基於 RPC 為基礎的服務發現。服務提供者使用原生 SDK、OpenAPI 或一個獨立的 Agent 註冊 Service 後，服務消費者可以使用 DNS 或 HTTP&API 尋找和發現服務。Nacos 提供對服務的即時的健康檢查，阻止向不健康的主機或服務實例發送請求。Nacos 支援傳輸層（PING 或 TCP）和應用層（如 HTTP、MySQL、使用者自訂）的健康檢查。對於複雜的雲環境和網路拓撲環境（如 VPC、邊緣網路等）中的服務的健康檢查，Nacos 提供了 Agent 上報模式和服務端主動檢測兩種健康檢查模式。Nacos 還提供了統一的健康檢查儀表板，可根據健康狀態管理服務的可用性及流量。

動態設定服務：動態設定服務有助以中心化、外部化和動態化的方式管理所有環境的應用設定和服務設定。動態設定消除了設定變更時重新部署應用和服務的需要，讓設定管理變得更加高效和敏捷。設定中心化管理讓實現無狀態服務變得更簡單，讓服務隨選彈性伸縮變得更容易。Nacos 提供了一個簡單好用的 UI 管理所有服務和應用的設定。Nacos 還提供包含設定版本追蹤、金絲雀發佈、一鍵回覆設定以及用戶端設定更新狀態追蹤在內的一系列開箱即用的設定管理特性，使人們可以更安全地在生產環境中管理設定變更和降低設定變更帶來的風險。

動態 DNS 服務：動態 DNS 服務支援權重路由，可更容易地實現中間層負載平衡、更靈活的路由策略、流量控制以及資料中心內網的簡單 DNS 解析服務。動態 DNS 服務可以實現以 DNS 協定為基礎的服務發現，消除耦合到廠商私有服務發現 API 上的風險。Nacos 還提供了一些簡單的 DNS API 管理服務的連結域名和可用的 IP:PORT 列表。

服務及其中繼資料管理：Nacos 從微服務平台建設的角度管理資料中心的所有服務及中繼資料，包含管理服務的描述、服務的生命週期、服務的靜態依賴分析、服務的健康狀態、服務的流量管理、路由及安全性原則、服務的 SLA 以及指標統計資料。

5.4.2 Kotlin 整合 Nacos 服務註冊

新增一個 Maven 子專案：chapter05-nacos，這是一個服務提供者，服務啟動後可以註冊到 Nacos。pom 檔案如下：

```
1.  <?xml version="1.0" encoding="UTF-8"?>
2.  <project xmlns="http://maven.apache.org/POM/4.0.0"
3.        xmlns:xsi="http://www.w3.org/2001/XMLSchema-instance"
4.        xsi:schemaLocation="http://maven.apache.org/POM/4.0.0 http://
    maven.apache.org/xsd/maven-4.0.0.xsd">
5.     <parent>
6.        <artifactId>kotlinspringboot</artifactId>
7.        <groupId>io.kang.kotlinspringboot</groupId>
8.        <version>0.0.1-SNAPSHOT</version>
```

```
9.         </parent>
10.        <modelVersion>4.0.0</modelVersion>
11.        <!-- 子專案名 -->
12.        <artifactId>chapter05-nacos</artifactId>
13.        <dependencies>
14.            <!-- Spring Cloud Nacos 相依套件 -->
15.            <dependency>
16.                <groupId>com.alibaba.cloud</groupId>
17.                <artifactId>spring-cloud-starter-alibaba-nacos-discovery
     </artifactId>
18.                <version>2.1.1.RELEASE</version>
19.            </dependency>
20.            <!-- Spring Boot Web 相依套件 -->
21.            <dependency>
22.                <groupId>org.springframework.boot</groupId>
23.                <artifactId>spring-boot-starter-web</artifactId>
24.                <version>2.2.1.RELEASE</version>
25.            </dependency>
26.            <dependency>
27.                <groupId>com.fasterxml.jackson.module</groupId>
28.                <artifactId>jackson-module-kotlin</artifactId>
29.            </dependency>
30.            <dependency>
31.                <groupId>org.jetbrains.kotlin</groupId>
32.                <artifactId>kotlin-reflect</artifactId>
33.            </dependency>
34.            <dependency>
35.                <groupId>org.jetbrains.kotlin</groupId>
36.                <artifactId>kotlin-stdlib-jdk8</artifactId>
37.            </dependency>
38.            <dependency>
39.                <groupId>org.jetbrains.kotlinx</groupId>
40.                <artifactId>kotlinx-coroutines-core</artifactId>
41.                <version>1.3.2</version>
42.            </dependency>
```

```
43.    </dependencies>
44.    <build>
45.        <sourceDirectory>${project.basedir}/src/main/kotlin
    </sourceDirectory>
46.        <testSourceDirectory>${project.basedir}/src/test/kotlin
    </testSourceDirectory>
47.        <plugins>
48.            <plugin>
49.                <groupId>org.springframework.boot</groupId>
50.                <artifactId>spring-boot-maven-plugin</artifactId>
51.            </plugin>
52.            <plugin>
53.                <groupId>org.jetbrains.kotlin</groupId>
54.                <artifactId>kotlin-maven-plugin</artifactId>
55.                <configuration>
56.                    <args>
57.                        <arg>-Xjsr305=strict</arg>
58.                    </args>
59.                    <compilerPlugins>
60.                        <plugin>spring</plugin>
61.                        <plugin>jpa</plugin>
62.                    </compilerPlugins>
63.                </configuration>
64.                <dependencies>
65.                    <dependency>
66.                        <groupId>org.jetbrains.kotlin</groupId>
67.                        <artifactId>kotlin-maven-allopen</artifactId>
68.                        <version>${kotlin.version}</version>
69.                    </dependency>
70.                    <dependency>
71.                        <groupId>org.jetbrains.kotlin</groupId>
72.                        <artifactId>kotlin-maven-noarg</artifactId>
73.                        <version>${kotlin.version}</version>
74.                    </dependency>
75.                </dependencies>
```

```
76.            </plugin>
77.          </plugins>
78.        </build>
79.  </project>
```

application.yml 檔案的內容如下:

```
1.  server:
2.    port: 8100                # 服務通訊埠編號
3.  spring:
4.    application:
5.      name: nacos-producer   # 服務名稱
6.    cloud:
7.      nacos:
8.        discovery:            # Nacos 服務中心位址
9.          server-addr: 127.0.0.1:8848
```

NacosApplication.kt 是一個啟動類別:

```
1.  // 開啟服務註冊註釋
2.  @SpringBootApplication
3.  @EnableDiscoveryClient
4.  class NacosApplication
5.  // 啟動函數
6.  fun main(args: Array<String>) {
7.      runApplication<NacosApplication>(*args)
8.  }
```

NacosController.kt 定義了一個服務介面:

```
1.  @RestController
2.  class NacosController {
3.      // 測試介面
4.      @GetMapping("hello/nacos")
5.      fun helloNacos(): String {
6.          return "Hello Nacos"
7.      }
8.  }
```

5.4.3 Kotlin 整合 OpenFeign 和 Ribbon 服務呼叫

新增一個 Maven 子專案：chapter05-nacos-consumer，這是一個服務消費方，採用 Feign 和 Ribbon 方式存取 chapter05-nacos 定義的服務介面。pom 檔案如下：

```
1.  <?xml version="1.0" encoding="UTF-8"?>
2.  <project xmlns="http://maven.apache.org/POM/4.0.0"
3.          xmlns:xsi="http://www.w3.org/2001/XMLSchema-instance"
4.          xsi:schemaLocation="http://maven.apache.org/POM/4.0.0 http://
    maven.apache.org/xsd/maven-4.0.0.xsd">
5.      <parent>
6.          <artifactId>kotlinspringboot</artifactId>
7.          <groupId>io.kang.kotlinspringboot</groupId>
8.          <version>0.0.1-SNAPSHOT</version>
9.      </parent>
10.     <modelVersion>4.0.0</modelVersion>
11.     <!-- 子專案名 -->
12.     <artifactId>chapter05-nacos-consumer</artifactId>
13.     <dependencies>
14.         <!-- Spring Cloud Nacos 相依套件 -->
15.         <dependency>
16.             <groupId>com.alibaba.cloud</groupId>
17.             <artifactId>spring-cloud-starter-alibaba-nacos-discovery
    </artifactId>
18.             <version>2.1.1.RELEASE</version>
19.         </dependency>
20.         <!-- Spring Boot Web 相依套件 -->
21.         <dependency>
22.             <groupId>org.springframework.boot</groupId>
23.             <artifactId>spring-boot-starter-web</artifactId>
24.             <version>2.2.1.RELEASE</version>
25.         </dependency>
26.         <!-- Spring Cloud OpenFeign 相依套件 -->
27.         <dependency>
```

```
28.            <groupId>org.springframework.cloud</groupId>
29.            <artifactId>spring-cloud-starter-openfeign</artifactId>
30.            <version>2.2.1.RELEASE</version>
31.        </dependency>
32.        <dependency>
33.            <groupId>com.fasterxml.jackson.module</groupId>
34.            <artifactId>jackson-module-kotlin</artifactId>
35.        </dependency>
36.        <dependency>
37.            <groupId>org.jetbrains.kotlin</groupId>
38.            <artifactId>kotlin-reflect</artifactId>
39.        </dependency>
40.        <dependency>
41.            <groupId>org.jetbrains.kotlin</groupId>
42.            <artifactId>kotlin-stdlib-jdk8</artifactId>
43.        </dependency>
44.        <dependency>
45.            <groupId>org.jetbrains.kotlinx</groupId>
46.            <artifactId>kotlinx-coroutines-core</artifactId>
47.            <version>1.3.2</version>
48.        </dependency>
49.    </dependencies>
50.    <build>
51.        <sourceDirectory>${project.basedir}/src/main/kotlin
    </sourceDirectory>
52.        <testSourceDirectory>${project.basedir}/src/test/kotlin
    </testSourceDirectory>
53.        <plugins>
54.            <plugin>
55.                <groupId>org.springframework.boot</groupId>
56.                <artifactId>spring-boot-maven-plugin</artifactId>
57.            </plugin>
58.            <plugin>
59.                <groupId>org.jetbrains.kotlin</groupId>
60.                <artifactId>kotlin-maven-plugin</artifactId>
```

```
61.                <configuration>
62.                    <args>
63.                        <arg>-Xjsr305=strict</arg>
64.                    </args>
65.                    <compilerPlugins>
66.                        <plugin>spring</plugin>
67.                        <plugin>jpa</plugin>
68.                    </compilerPlugins>
69.                </configuration>
70.                <dependencies>
71.                    <dependency>
72.                        <groupId>org.jetbrains.kotlin</groupId>
73.                        <artifactId>kotlin-maven-allopen</artifactId>
74.                        <version>${kotlin.version}</version>
75.                    </dependency>
76.                    <dependency>
77.                        <groupId>org.jetbrains.kotlin</groupId>
78.                        <artifactId>kotlin-maven-noarg</artifactId>
79.                        <version>${kotlin.version}</version>
80.                    </dependency>
81.                </dependencies>
82.            </plugin>
83.        </plugins>
84.    </build>
85. </project>
```

application.yml 檔案的內容如下：

```
1. server:
2.   port: 8006                   # 應用通訊埠編號
3. spring:
4.   application:
5.     name: nacos-consumer   # 服務名稱
6.   cloud:
7.     nacos:
8.       discovery:              # Nacos 註冊中心位址
9.         server-addr: 127.0.0.1:8848
```

NacosConsumerApplication.kt 定義了一個啟動類別：

```
1.  // 開啟服務註冊註釋，開啟 Feign 註釋
2.  @SpringBootApplication
3.  @EnableFeignClients
4.  @EnableDiscoveryClient
5.  class NacosConsumerApplication {
6.      @Bean
7.      @LoadBalanced
8.      fun restTemplate(): RestTemplate {
9.          return RestTemplate()
10.     }
11. }
12. // 啟動函數
13. fun main(args: Array<String>) {
14.     runApplication<NacosConsumerApplication>(*args)
15. }
```

FeignService.kt 定義了一個 Feign 介面：

```
1.  @FeignClient(value = "nacos-producer")
2.  @Component
3.  interface FeignService {
4.      // 測試介面，使用 Feign 方式呼叫
5.      @GetMapping("hello/nacos")
6.      fun helloNacos(): String
7.  }
```

NacosConsumerController.kt 定義了兩個測試介面，分別用 Ribbon 和 Feign
方式呼叫服務介面：

```
1.  @RestController
2.  class NacosConsumerController {
3.      @Autowired
4.      lateinit var restTemplate: RestTemplate
5.      @Autowired
6.      lateinit var feignService: FeignService
7.      // 測試介面，使用 Ribbon 方式呼叫
```

```
8.      @GetMapping("ribbon/hello/nacos")
9.      fun ribbonHelloNacos(): String? {
10.         return restTemplate.getForObject("http://nacos-producer/hello/
    nacos", String::class.java)
11.     }
12.     // 測試介面，使用 Feign 方式呼叫
13.     @GetMapping("feign/hello/nacos")
14.     fun feignHelloNacos(): String {
15.         return feignService.helloNacos()
16.     }
17. }
```

使用 Ribbon 方式透過 "ribbon/hello/nacos" 或使用 Feign 方式透過 "feign/hello/nacos" 都可以存取到 "hello/nacos" 介面。

Nacos 監控頁面如圖 5.6 所示。

圖 5.6 Nacos 監控介面

5.5 小結

本章介紹了 Kotlin 應用於微服務註冊中心的相關內容，透過範例介紹了服務註冊和呼叫方法。服務註冊中心對於微服務系統很重要，可影響服務的註冊、發現、治理。使用服務註冊中心，可將單體應用拆分為微服務，微服務元件透過服務註冊中心通訊。

Chapter 06

Kotlin 應用於微服務
設定中心

設定中心可以儲存微服務系統的設定，當設定更新時，可以下發到微服務系統。本章將介紹四種設定中心：Spring Cloud Config、Apollo、Nacos 和 Consul。本章將使用範例展示 Kotlin 整合微服務設定中心的實際方法。

6.1 Spring Cloud Config

Spring Cloud Config 是 Spring Cloud 家族的中心元件，負責設定檔的統一管理及即時更新。本節主要介紹 Spring Cloud Config 作為 Kotlin 開發的微服務的設定中心的相關知識。

6.1.1 Spring Cloud Config 介紹

Spring Cloud Config 支援將設定檔放在設定服務的記憶體中（即本機），也支援放在遠端 Git 倉庫中。在 Spring Cloud Config 元件中有兩種角色，一種是 config server，另一種是 config client。可以透過 RESTful 介面存取設定檔、請求位址和儲存資源檔對映，如下所示：

```
/{application}/{profile}[/{label}]
/{application}-{profile}.yml
/{label}/{application}-{profile}.yml
/{application}-{profile}.properties
/{label}/{application}-{profile}.properties
```

application 是應用的名稱，profile 區分開發環境、測試環境、生產環境設定檔，label 是設定的分支標籤，用於版本管理。Spring Boot 支援 yml 和 properties 兩種格式的設定檔。

服務實例都將從設定中心讀取檔案，可以將設定中心做成一個微服務，將其叢集化，進一步達到高可用，如圖 6.1 所示。

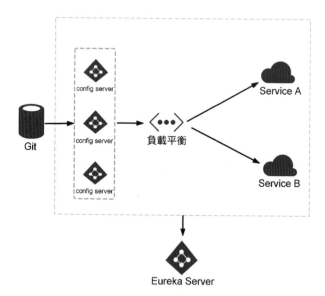

圖 6.1 服務設定中心示意圖

Spring Cloud Config 借 助 Spring Cloud Bus 實 現 設 定 的 熱 載 入。Spring Cloud Bus 會對外提供一個 HTTP 介面，即 /bus-refresh。將這個介面設定到遠端的 Git 的 webhook 上，當 Git 上的檔案內容發生變動時，Git 會自動呼叫 /bus-refresh 介面。Bus 就會通知 config server，config server 會發佈

更新訊息到訊息匯流排的訊息佇列中，其他服務訂閱到該訊息就會進行更新，進一步實現整個微服務進行自動更新。

6.1.2 Kotlin 整合 Spring Cloud Config

新增一個 Maven 子專案：chapter06-springcloud-config，這是一個 config server。pom 檔案如下：

```
1.  <?xml version="1.0" encoding="UTF-8"?>
2.  <project xmlns="http://maven.apache.org/POM/4.0.0"
3.          xmlns:xsi="http://www.w3.org/2001/XMLSchema-instance"
4.          xsi:schemaLocation="http://maven.apache.org/POM/4.0.0 http://
    maven.apache.org/xsd/maven-4.0.0.xsd">
5.      <parent>
6.          <artifactId>kotlinspringboot</artifactId>
7.          <groupId>io.kang.kotlinspringboot</groupId>
8.          <version>0.0.1-SNAPSHOT</version>
9.      </parent>
10.     <modelVersion>4.0.0</modelVersion>
11.     <!-- 子專案名 -->
12.     <artifactId>chapter06-springcloud-config</artifactId>
13.     <dependencies>
14.         <!-- Spring Cloud Config Server 相依套件 -->
15.         <dependency>
16.             <groupId>org.springframework.cloud</groupId>
17.             <artifactId>spring-cloud-config-server</artifactId>
18.             <version>2.2.1.RELEASE</version>
19.         </dependency>
20.         <dependency>
21.             <groupId>com.fasterxml.jackson.module</groupId>
22.             <artifactId>jackson-module-kotlin</artifactId>
23.         </dependency>
24.         <dependency>
25.             <groupId>org.jetbrains.kotlin</groupId>
```

```
26.              <artifactId>kotlin-reflect</artifactId>
27.          </dependency>
28.          <dependency>
29.              <groupId>org.jetbrains.kotlin</groupId>
30.              <artifactId>kotlin-stdlib-jdk8</artifactId>
31.          </dependency>
32.          <dependency>
33.              <groupId>org.jetbrains.kotlinx</groupId>
34.              <artifactId>kotlinx-coroutines-core</artifactId>
35.              <version>1.3.2</version>
36.          </dependency>
37.      </dependencies>
38.      <build>
39.          <sourceDirectory>${project.basedir}/src/main/kotlin
    </sourceDirectory>
40.          <testSourceDirectory>${project.basedir}/src/test/kotlin
    </testSourceDirectory>
41.          <plugins>
42.              <plugin>
43.                  <groupId>org.springframework.boot</groupId>
44.                  <artifactId>spring-boot-maven-plugin</artifactId>
45.              </plugin>
46.              <plugin>
47.                  <groupId>org.jetbrains.kotlin</groupId>
48.                  <artifactId>kotlin-maven-plugin</artifactId>
49.                  <configuration>
50.                      <args>
51.                          <arg>-Xjsr305=strict</arg>
52.                      </args>
53.                      <compilerPlugins>
54.                          <plugin>spring</plugin>
55.                          <plugin>jpa</plugin>
56.                      </compilerPlugins>
57.                  </configuration>
58.                  <dependencies>
59.                      <dependency>
```

```
60.                    <groupId>org.jetbrains.kotlin</groupId>
61.                    <artifactId>kotlin-maven-allopen</artifactId>
62.                    <version>${kotlin.version}</version>
63.                  </dependency>
64.                  <dependency>
65.                    <groupId>org.jetbrains.kotlin</groupId>
66.                    <artifactId>kotlin-maven-noarg</artifactId>
67.                    <version>${kotlin.version}</version>
68.                  </dependency>
69.                </dependencies>
70.              </plugin>
71.            </plugins>
72.          </build>
73.        </project>
```

application.yml 檔案的內容如下：

```
1.   spring:
2.     application:
3.       name: configserver              # 應用名
4.     cloud:
5.       config:
6.         server:
7.           git:
8.             uri: https://github.com/dutyk/kotlinmicroservice   #Git 倉庫位址
9.             username:                  #Git 使用者名稱
10.            password:                  #Git 密碼
11.            default-label: master  # 設定檔分支
12.            search-paths: config   # 設定檔所在根目錄
13.  server:
14.    port: 8060                         # 應用通訊埠編號
```

在 GitHub 上上傳一個設定檔：springcloudconfig-dev.yml，定義如下：

```
1.   data:
2.     env: dev
3.     user:
```

```
4.    username: user_dev01
5.    password: password_dev
```

ConfigApplication.kt 定義了一個啟動類別，啟動一個設定中心：

```
1.  // 開啟 config server 註釋
2.  @SpringBootApplication
3.  @EnableConfigServer
4.  class ConfigApplication
5.  // 啟動函數
6.  fun main(args: Array<String>) {
7.      runApplication<ConfigApplication>(*args)
8.  }
```

新增一個 Maven 子專案：chapter06-springcloud-config-client，這是一個用戶端，從 config server 取得設定。pom 檔案如下：

```
1.  <?xml version="1.0" encoding="UTF-8"?>
2.  <project xmlns="http://maven.apache.org/POM/4.0.0"
3.          xmlns:xsi="http://www.w3.org/2001/XMLSchema-instance"
4.          xsi:schemaLocation="http://maven.apache.org/POM/4.0.0 http://
    maven.apache.org/xsd/maven-4.0.0.xsd">
5.      <parent>
6.          <artifactId>kotlinspringboot</artifactId>
7.          <groupId>io.kang.kotlinspringboot</groupId>
8.          <version>0.0.1-SNAPSHOT</version>
9.      </parent>
10.     <modelVersion>4.0.0</modelVersion>
11.     <!-- 子專案名 -->
12.     <artifactId>chapter06-springcloud-config-client</artifactId>
13.     <dependencies>
14.         <!-- Spring Cloud Config 相依套件 -->
15.         <dependency>
16.             <groupId>org.springframework.cloud</groupId>
17.             <artifactId>spring-cloud-starter-config</artifactId>
18.             <version>2.2.1.RELEASE</version>
```

```
19.        </dependency>
20.        <!-- Spring Boot Web 相依套件 -->
21.        <dependency>
22.            <groupId>org.springframework.boot</groupId>
23.            <artifactId>spring-boot-starter-web</artifactId>
24.            <version>2.2.1.RELEASE</version>
25.        </dependency>
26.        <!-- Spring Boot Actuator 相依套件 -->
27.        <dependency>
28.            <groupId>org.springframework.boot</groupId>
29.            <artifactId>spring-boot-starter-actuator</artifactId>
30.            <version>2.2.1.RELEASE</version>
31.        </dependency>
32.        <dependency>
33.            <groupId>com.fasterxml.jackson.module</groupId>
34.            <artifactId>jackson-module-kotlin</artifactId>
35.        </dependency>
36.        <dependency>
37.            <groupId>org.jetbrains.kotlin</groupId>
38.            <artifactId>kotlin-reflect</artifactId>
39.        </dependency>
40.        <dependency>
41.            <groupId>org.jetbrains.kotlin</groupId>
42.            <artifactId>kotlin-stdlib-jdk8</artifactId>
43.        </dependency>
44.        <dependency>
45.            <groupId>org.jetbrains.kotlinx</groupId>
46.            <artifactId>kotlinx-coroutines-core</artifactId>
47.            <version>1.3.2</version>
48.        </dependency>
49.    </dependencies>
50.
51.    <build>
52.        <sourceDirectory>${project.basedir}/src/main/kotlin
    </sourceDirectory>
```

```
53.          <testSourceDirectory>${project.basedir}/src/test/kotlin
      </testSourceDirectory>
54.          <plugins>
55.             <plugin>
56.                 <groupId>org.springframework.boot</groupId>
57.                 <artifactId>spring-boot-maven-plugin</artifactId>
58.             </plugin>
59.             <plugin>
60.                 <groupId>org.jetbrains.kotlin</groupId>
61.                 <artifactId>kotlin-maven-plugin</artifactId>
62.                 <configuration>
63.                     <args>
64.                         <arg>-Xjsr305=strict</arg>
65.                     </args>
66.                     <compilerPlugins>
67.                         <plugin>spring</plugin>
68.                         <plugin>jpa</plugin>
69.                     </compilerPlugins>
70.                 </configuration>
71.                 <dependencies>
72.                     <dependency>
73.                         <groupId>org.jetbrains.kotlin</groupId>
74.                         <artifactId>kotlin-maven-allopen</artifactId>
75.                         <version>${kotlin.version}</version>
76.                     </dependency>
77.                     <dependency>
78.                         <groupId>org.jetbrains.kotlin</groupId>
79.                         <artifactId>kotlin-maven-noarg</artifactId>
80.                         <version>${kotlin.version}</version>
81.                     </dependency>
82.                 </dependencies>
83.             </plugin>
84.         </plugins>
85.     </build>
86. </project>
```

application.yml 檔案的內容如下：

```
1.  management:
2.    endpoints:
3.      web:
4.        exposure:
5.          include: refresh,health,info    # 曝露監控介面
```

同時，還定義了一個 bootstrap.yml：

```
1.  server:
2.    port: 8061                      # 應用通訊埠
3.  spring:
4.    application:
5.      name: springcloudconfig       # 應用名稱
6.    cloud:
7.      config:
8.        label: master               # 取得 master 分支
9.        profile: dev                # 取得 application-dev.yml 檔案設定
10.       uri: http://localhost:8060  # 設定中心位址
```

ClientApplication.kt 定義了一個啟動類別，啟動一個 Spring Boot Web 應用：

```
1.  @SpringBootApplication
2.  class ClientApplication
3.  // 啟動函數
4.  fun main(args: Array<String>) {
5.      runApplication<ClientApplication>(*args)
6.  }
```

GitConfig.kt 定義了一些屬性及對應的值，這些屬性從設定中心取得：

```
1.  @Component
2.  // 開啟更新
3.  @RefreshScope
4.  // 和設定檔定義的值一一對應
```

```
5.  data class GitConfig(
6.         @Value("\${data.env}")
7.         val env: String,
8.         @Value("\${data.user.username}")
9.         val username: String,
10.        @Value("\${data.user.password}")
11.        val password: String
12. )
```

ConfigController.kt 定義了一個測試介面：

```
1.  @RestController
2.  class ConfigController {
3.      @Autowired
4.      lateinit var gitConfig: GitConfig
5.      // 測試介面，用於取得設定
6.      @GetMapping("/config")
7.      fun getConfig(): String {
8.          return gitConfig.toString()
9.      }
10. }
```

存取 "/config" 介面，可以取得到 env、username、password 這幾個設定項
目。修改 env 的值，提交到 Git 倉庫，然後呼叫 "actuator/refresh" 更新設
定，再存取 "/config"，可取得最新的值。

6.2 Apollo 設定中心

Apollo 是攜程研發的設定中心，能夠集中管理應用程式在開發、測試、生
產環境中的設定資訊。Apollo 提供了一個統一介面集中管理設定，支援多
環境、多資料中心設定及許可權管理等特性。本節介紹使用 Apollo 作為
Kotlin 開發的微服務的設定中心的相關知識。

6.2.1 Apollo 介紹

Apollo 中的設定是獨立於程式的唯讀變數，其伴隨應用的整個生命週期。其中的設定可以有多種載入方式，常見的有程式內部的強制寫入、設定檔、環境變數、啟動參數、基於資料庫等。設定需要治理，需要進行許可權控制，以管理不同環境和叢集的設定。Apollo 從 4 個維度管理 key-value 的設定。

- application（應用）：每個應用都需要有唯一的身份標識 appId。
- environment（環境）：環境預設是透過讀取機器上的設定（server. properties 中的 env 屬性）指定的，也支援執行時期透過系統內容等指定。
- cluster（叢集）：叢集預設是透過讀取機器上的設定（server.properties 中的 idc 屬性）指定的，也支援執行時期透過系統內容指定。
- namespace（命名空間）：一個應用的不同設定的分組。可以簡單地把 namespace 想像成檔案，不同類型的設定儲存在不同的檔案中，可以直接讀取，也可繼承公共設定。同一份程式部署在不同的叢集，可以有不同的設定，例如 Zookeeper 的位址等。透過命名空間可以很方便地支援多個不同應用共用同一份設定，同時還允許應用對共用的設定進行覆蓋。

使用者在 Apollo 中修改完設定並發佈後，用戶端能即時（1 秒）接收到最新的設定，並通知到應用程式。所有的設定發佈都有版本概念，進一步可以方便地支援設定的回覆。Apollo 還支援設定的灰階發佈，例如發佈後，只對部分應用實例生效，等觀察一段時間沒問題後再推給所有應用實例。Apollo 對應用和設定的管理有完整的許可權管理機制，對設定的管理還分為編輯和發佈兩個環節，可減少人為造成的錯誤。

Apollo 目前唯一的外部依賴是 MySQL，所以部署非常簡單，只要安裝好 Java 和 MySQL 就可以讓 Apollo 執行起來。Apollo 還提供了包裝指令稿，一鍵就可以產生所有需要的安裝套件，並且支援自訂執行時期參數。

Apollo 的核心元件有以下幾個。

- ConfigService：提供設定取得介面，提供設定發送介面，服務於 Apollo 用戶端。
- AdminService：提供設定管理介面，提供設定修改發佈介面，服務於管理介面 Portal。
- Client：為應用取得設定，支援即時更新，透過 MetaServer 取得 Config Service 的服務清單，使用用戶端軟負載 SLB 方式呼叫 ConfigService。
- Portal：設定管理介面，透過 MetaServer 取得 AdminService 的服務清單，使用用戶端軟負載 SLB 方式呼叫 AdminService。
- Eureka：用於服務發現和註冊，ConfigService 和 AdminService 在 Eureka 註冊實例並定期上報心跳，Eureka 和 ConfigService 在一起部署。
- MetaServer：Portal 透過域名存取 MetaServer 取得 AdminService 的位址清單，Client 透過域名存取 MetaServer 取得 ConfigService 的位址清單。MetaServer 相當於一個 Eureka Proxy，MetaServer 和 ConfigService 在一起部署。
- NginxLB：和網域名稱系統配合，協助 Portal 存取 MetaServer 取得 AdminService 的位址清單；和網域名稱系統配合，協助 Client 存取 MetaServer 取得 ConfigService 位址清單；和網域名稱系統配合，協助使用者存取 Portal 進行設定管理。

6.2.2 Kotlin 整合 Apollo

啟動 Apollo 服務，在 Apollo 中增加如圖 6.2 所示的設定。

timeout	100	sample timeout配置		2020-02-11 13:00:37	
data.env	dev		apollo	2020-02-11 14:16:55	
data.user.username	dev01		apollo	2020-02-11 15:26:09	
data.user.password	dev01		apollo	2020-02-11 14:17:16	

圖 6.2 Apollo 的設定介面

新增一個 Maven 子專案：chapter06-apollo-config，從 Apollo 取得設定。
pom 檔案如下：

```
1.  <?xml version="1.0" encoding="UTF-8"?>
2.  <project xmlns="http://maven.apache.org/POM/4.0.0"
3.         xmlns:xsi="http://www.w3.org/2001/XMLSchema-instance"
4.         xsi:schemaLocation="http://maven.apache.org/POM/4.0.0 http://
    maven.apache.org/xsd/maven-4.0.0.xsd">
5.      <parent>
6.          <artifactId>kotlinspringboot</artifactId>
7.          <groupId>io.kang.kotlinspringboot</groupId>
8.          <version>0.0.1-SNAPSHOT</version>
9.      </parent>
10.     <modelVersion>4.0.0</modelVersion>
11.     <!-- 子專案名 -->
12.     <artifactId>chapter06-apollo-config</artifactId>
13.     <dependencies>
14.         <!-- Apollo 用戶端相依套件 -->
15.         <dependency>
16.             <groupId>com.ctrip.framework.apollo</groupId>
17.             <artifactId>apollo-client</artifactId>
18.             <version>1.5.1</version>
19.         </dependency>
20.         <!-- Apollo 核心相依套件 -->
21.         <dependency>
22.             <groupId>com.ctrip.framework.apollo</groupId>
23.             <artifactId>apollo-core</artifactId>
24.             <version>1.5.1</version>
25.         </dependency>
26.         <!-- Spring Boot Web 相依套件 -->
27.         <dependency>
28.             <groupId>org.springframework.boot</groupId>
29.             <artifactId>spring-boot-starter-web</artifactId>
30.             <version>2.2.1.RELEASE</version>
31.         </dependency>
```

```
32.          <dependency>
33.              <groupId>com.fasterxml.jackson.module</groupId>
34.              <artifactId>jackson-module-kotlin</artifactId>
35.          </dependency>
36.          <dependency>
37.              <groupId>org.jetbrains.kotlin</groupId>
38.              <artifactId>kotlin-reflect</artifactId>
39.          </dependency>
40.          <dependency>
41.              <groupId>org.jetbrains.kotlin</groupId>
42.              <artifactId>kotlin-stdlib-jdk8</artifactId>
43.          </dependency>
44.          <dependency>
45.              <groupId>org.jetbrains.kotlinx</groupId>
46.              <artifactId>kotlinx-coroutines-core</artifactId>
47.              <version>1.3.2</version>
48.          </dependency>
49.      </dependencies>
50.      <build>
51.          <sourceDirectory>${project.basedir}/src/main/kotlin
     </sourceDirectory>
52.          <testSourceDirectory>${project.basedir}/src/test/kotlin
     </testSourceDirectory>
53.          <plugins>
54.              <plugin>
55.                  <groupId>org.springframework.boot</groupId>
56.                  <artifactId>spring-boot-maven-plugin</artifactId>
57.              </plugin>
58.              <plugin>
59.                  <groupId>org.jetbrains.kotlin</groupId>
60.                  <artifactId>kotlin-maven-plugin</artifactId>
61.                  <configuration>
62.                      <args>
63.                          <arg>-Xjsr305=strict</arg>
64.                      </args>
```

```
65.                    <compilerPlugins>
66.                        <plugin>spring</plugin>
67.                        <plugin>jpa</plugin>
68.                    </compilerPlugins>
69.                </configuration>
70.                <dependencies>
71.                    <dependency>
72.                        <groupId>org.jetbrains.kotlin</groupId>
73.                        <artifactId>kotlin-maven-allopen</artifactId>
74.                        <version>${kotlin.version}</version>
75.                    </dependency>
76.                    <dependency>
77.                        <groupId>org.jetbrains.kotlin</groupId>
78.                        <artifactId>kotlin-maven-noarg</artifactId>
79.                        <version>${kotlin.version}</version>
80.                    </dependency>
81.                </dependencies>
82.            </plugin>
83.        </plugins>
84.    </build>
85. </project>
```

application.yml 檔案的內容如下：

```
1.  app:
2.    id: SampleApp              # app id 名
3.  apollo:
4.    meta: http://127.0.0.1:8080    # Apollo 設定中心位址
5.  server:
6.    port: 8062                 # 應用通訊埠編號
```

ApolloApplication.kt 是一個啟動類別，啟動一個 Spring Boot Web 應用：

```
1.  // 開啟 Apollo config 註釋
2.  @SpringBootApplication
3.  @EnableApolloConfig
```

```
4.  class ApolloApplication
5.  // 啟動函數
6.  fun main(args: Array<String>) {
7.      System.setProperty("env", "dev")
8.      runApplication<ApolloApplication>(*args)
9.  }
```

ApolloConfig.kt 定義了 Apollo 中的設定項目並監聽設定更新：

```
1.  // 設定實體類別
2.  @Configuration
3.  @ConfigurationProperties(prefix = "data")
4.  class ApolloConfig {
5.      var env: String? = null
6.      var user: User? = null
7.      // 監聽函數，當 username 值更新時觸發該函數
8.      @ApolloConfigChangeListener
9.      fun configChangeHandlerUserName(configChangeEvent: ConfigChangeEvent) {
    if(configChangeEvent.isChanged("data.user.username")) {
10.             user?.username = configChangeEvent.getChange("data.user.
    username").newValue
11.             println("${user?.username} is change")
12.         }
13.     }
14. }
15. // User 實體類別
16. class User{
17.     var username: String? = null
18.     var password: String? = null
19.     override fun toString(): String {
20.         return "${username},${password}"
21.     }
22. }
```

ApolloController.kt 定義了一個測試介面：

```
1.  @RestController
2.  class ApolloController {
```

```
3.      @Autowired
4.      lateinit var apolloConfig: ApolloConfig
5.      // 測試介面，測試從 Apollo 讀取設定值
6.      @GetMapping("config")
7.      fun getApolloConfig(): String {
8.          return "${apolloConfig.env},${apolloConfig.user}"
9.      }
10. }
```

存取 "config" 介面，可以取得 env 和 user 的值。修改 env 的值，發佈
Apollo 設定，再次存取 "config" 介面，可以取得最新的 env 值。

6.3 Nacos 設定中心

本節將介紹使用 Nacos 作為 Kotlin 開發的微服務的設定中心的相關知
識。在 Nacos 的設定介面中新增了兩個 Data ID：nacos-config-dev.yaml 和
nacos-config.yaml，並新增了一些設定項目，如圖 6.3 和圖 6.4 所示。

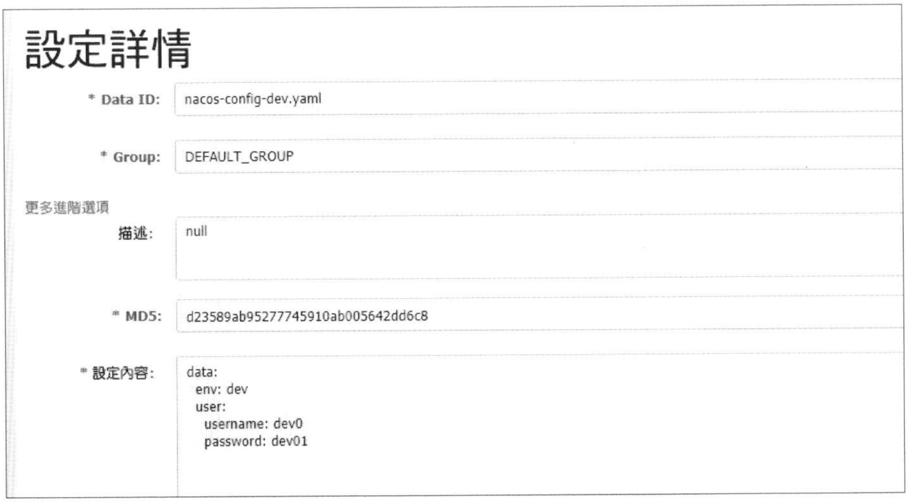

圖 6.3 nacos-config-dev.yaml 設定詳情

圖 6.4　nacos-config.yaml 設定詳情

新增一個 Maven 子專案：chapter06-nacos-config，這是一個用戶端，可以
從 Nacos 取得設定。pom 檔案如下：

```xml
1.  <?xml version="1.0" encoding="UTF-8"?>
2.  <project xmlns="http://maven.apache.org/POM/4.0.0"
3.          xmlns:xsi="http://www.w3.org/2001/XMLSchema-instance"
4.          xsi:schemaLocation="http://maven.apache.org/POM/4.0.0 http://
    maven.apache.org/xsd/maven-4.0.0.xsd">
5.      <parent>
6.          <artifactId>kotlinspringboot</artifactId>
7.          <groupId>io.kang.kotlinspringboot</groupId>
8.          <version>0.0.1-SNAPSHOT</version>
9.      </parent>
10.     <modelVersion>4.0.0</modelVersion>
11.     <!-- 子專案名 -->
12.     <artifactId>chapter06-nacos-config</artifactId>
13.     <dependencies>
14.         <!-- Spring Cloud Nacos 相依套件 -->
15.         <dependency>
16.             <groupId>com.alibaba.cloud</groupId>
17.             <artifactId>spring-cloud-starter-alibaba-nacos-config
```

```xml
          </artifactId>
18.             <version>2.2.0.RELEASE</version>
19.         </dependency>
20.         <!-- Spring Boot Web 相依套件 -->
21.         <dependency>
22.             <groupId>org.springframework.boot</groupId>
23.             <artifactId>spring-boot-starter-web</artifactId>
24.             <version>2.2.1.RELEASE</version>
25.         </dependency>
26.         <dependency>
27.             <groupId>com.fasterxml.jackson.module</groupId>
28.             <artifactId>jackson-module-kotlin</artifactId>
29.         </dependency>
30.         <dependency>
31.             <groupId>org.jetbrains.kotlin</groupId>
32.             <artifactId>kotlin-reflect</artifactId>
33.         </dependency>
34.         <dependency>
35.             <groupId>org.jetbrains.kotlin</groupId>
36.             <artifactId>kotlin-stdlib-jdk8</artifactId>
37.         </dependency>
38.         <dependency>
39.             <groupId>org.jetbrains.kotlinx</groupId>
40.             <artifactId>kotlinx-coroutines-core</artifactId>
41.             <version>1.3.2</version>
42.         </dependency>
43.     </dependencies>
44.     <build>
45.         <sourceDirectory>${project.basedir}/src/main/kotlin
    </sourceDirectory>
46.         <testSourceDirectory>${project.basedir}/src/test/kotlin
    </testSourceDirectory>
47.         <plugins>
48.             <plugin>
49.                 <groupId>org.springframework.boot</groupId>
```

```
50.              <artifactId>spring-boot-maven-plugin</artifactId>
51.          </plugin>
52.          <plugin>
53.              <groupId>org.jetbrains.kotlin</groupId>
54.              <artifactId>kotlin-maven-plugin</artifactId>
55.              <configuration>
56.                  <args>
57.                      <arg>-Xjsr305=strict</arg>
58.                  </args>
59.                  <compilerPlugins>
60.                      <plugin>spring</plugin>
61.                      <plugin>jpa</plugin>
62.                  </compilerPlugins>
63.              </configuration>
64.              <dependencies>
65.                  <dependency>
66.                      <groupId>org.jetbrains.kotlin</groupId>
67.                      <artifactId>kotlin-maven-allopen</artifactId>
68.                      <version>${kotlin.version}</version>
69.                  </dependency>
70.                  <dependency>
71.                      <groupId>org.jetbrains.kotlin</groupId>
72.                      <artifactId>kotlin-maven-noarg</artifactId>
73.                      <version>${kotlin.version}</version>
74.                  </dependency>
75.              </dependencies>
76.          </plugin>
77.      </plugins>
78.  </build>
79. </project>
```

bootstrap.yml 檔案的內容如下所示，其中指定了設定中心的位址和設定檔的副檔名。Nacos 的 data-id 名稱由 spring.application.name 和 spring.profiles.active 組成，副檔名是 yaml。此外，還可以用 extension-configs 指定 data-id：

```
1.  spring:
2.    cloud:
3.      nacos:
4.        config:
5.          server-addr: 127.0.0.1:8848      # Nacos 設定中心的位址
6.          file-extension: yaml             # 設定檔採用 yaml 格式
7.          extension-configs:
8.            -
9.              data-id: nacos-config.yaml   # data-id 定義
10.             group: default               # group 定義
11.             refresh: true                # 設定自動更新
12.     profiles:
13.       active: dev                        # 讀取 dev 結尾的設定檔
14.     application:
15.       name: nacos-config                 # 應用名
16. server:
17.   port: 8063                             # 應用通訊埠編號
```

NacosApplication.kt 是一個啟動類別，啟動了一個 Spring Boot Web 應用：

```
1.  @SpringBootApplication
2.  class NacosApplication
3.  // 啟動函數
4.  fun main(args: Array<String>) {
5.      runApplication<NacosApplication>(*args)
6.  }
```

NacosConfig.kt 定義了 nacos-config-dev.yaml 的設定項目。當更新 Nacos 設定時，使用 @ConfigurationProperties 註釋，NacosConfig 這個類別的屬性值也會更新：

```
1.  // 實體類別，對應 nacos 的設定項目
2.  @Component
3.  @ConfigurationProperties(prefix = "data")
4.  class NacosConfig {
5.      var env: String? = null
6.      var user: User? = null
7.  }
```

```
8.    // User 實體類別
9.    class User {
10.       var username: String? = null
11.       var password: String? = null
12.       override fun toString(): String {
13.           return "${username},${password}"
14.       }
15. }
```

NacosConfig1.kt 定義了 nacos-config.yaml 的設定項目。使用 @Value 註
釋，Nacos 的設定更新，NacosConfig1 的屬性值不會更新，重新啟動應用
才會更新：

```
1.    // 實體類別，對應 Nacos 設定項目
2.    @Configuration
3.    class NacosConfig1 {
4.       @Value(value = "\${data1.env}")
5.       var env: String? = null
6.       @Value(value = "\${data1.username}")
7.       var username: String? = null
8.       @Value(value = "\${data1.password}")
9.       var password: String? = null
10. }
```

NacosController.kt 定義了兩個測試介面：

```
1.    @RestController
2.    class NacosController {
3.       @Autowired
4.       lateinit var nacosConfig: NacosConfig
5.       @Autowired
6.       lateinit var nacosConfig1: NacosConfig1
7.       // 測試介面，讀取 Nacos 設定項目
8.       @GetMapping("config")
9.       fun getNacosConfig(): String {
10.          return "${nacosConfig.env}-${nacosConfig.user}"
11.       }
12.       // 測試介面，讀取 Nacos 設定項目
```

```
13.    @GetMapping("config1")
14.    fun getNacosConfig1(): String {
15.        return "${nacosConfig1.env}-${nacosConfig1.username}-
   ${nacosConfig1.password}"
16.    }
17. }
```

存取 "config" 介面可以取得 nacos-config-dev.yaml 的設定項目。更新 Nacos 的設定，並發佈，可以自動更新設定，再次存取 "config" 介面，可以取得最新的設定。存取 "config1" 可以取得 nacos-config.yaml 對應的設定。

6.4 Consul 設定中心

本節介紹使用 Consul 作為 Kotlin 開發的微服務的設定中心的相關知識。在 Consul 中設定如圖 6.5 所示的屬性。

圖 6.5　Consul 設定介面

新增一個 Maven 子專案：chapter06-consul-config，它從 Consul 讀取設定，pom 檔案如下：

```
1.  <?xml version="1.0" encoding="UTF-8"?>
2.  <project xmlns="http://maven.apache.org/POM/4.0.0"
3.          xmlns:xsi="http://www.w3.org/2001/XMLSchema-instance"
4.          xsi:schemaLocation="http://maven.apache.org/POM/4.0.0 http://
    maven.apache.org/xsd/maven-4.0.0.xsd">
5.      <parent>
6.          <artifactId>kotlinspringboot</artifactId>
7.          <groupId>io.kang.kotlinspringboot</groupId>
8.          <version>0.0.1-SNAPSHOT</version>
9.      </parent>
10.     <modelVersion>4.0.0</modelVersion>
11.     <!-- 子專案名 -->
12.     <artifactId>chapter06-consul-config</artifactId>
13.     <dependencies>
14.         <!-- Spring Cloud Consul Config 相依套件 -->
15.         <dependency>
16.             <groupId>org.springframework.cloud</groupId>
17.             <artifactId>spring-cloud-starter-consul-config</artifactId>
18.             <version>2.2.1.RELEASE</version>
19.         </dependency>
20.         <!-- Spring Boot Web 相依套件 -->
21.         <dependency>
22.             <groupId>org.springframework.boot</groupId>
23.             <artifactId>spring-boot-starter-web</artifactId>
24.             <version>2.2.1.RELEASE</version>
25.         </dependency>
26.         <!-- Spring Boot Actuator 相依套件 -->
27.         <dependency>
28.             <groupId>org.springframework.boot</groupId>
29.             <artifactId>spring-boot-starter-actuator</artifactId>
30.             <version>2.2.1.RELEASE</version>
31.         </dependency>
```

```
32.        <dependency>
33.            <groupId>com.fasterxml.jackson.module</groupId>
34.            <artifactId>jackson-module-kotlin</artifactId>
35.        </dependency>
36.        <dependency>
37.            <groupId>org.jetbrains.kotlin</groupId>
38.            <artifactId>kotlin-reflect</artifactId>
39.        </dependency>
40.        <dependency>
41.            <groupId>org.jetbrains.kotlin</groupId>
42.            <artifactId>kotlin-stdlib-jdk8</artifactId>
43.        </dependency>
44.        <dependency>
45.            <groupId>org.jetbrains.kotlinx</groupId>
46.            <artifactId>kotlinx-coroutines-core</artifactId>
47.            <version>1.3.2</version>
48.        </dependency>
49.    </dependencies>
50.    <build>
51.        <sourceDirectory>${project.basedir}/src/main/kotlin
    </sourceDirectory>
52.        <testSourceDirectory>${project.basedir}/src/test/kotlin
    </testSourceDirectory>
53.        <plugins>
54.            <plugin>
55.                <groupId>org.springframework.boot</groupId>
56.                <artifactId>spring-boot-maven-plugin</artifactId>
57.            </plugin>
58.            <plugin>
59.                <groupId>org.jetbrains.kotlin</groupId>
60.                <artifactId>kotlin-maven-plugin</artifactId>
61.                <configuration>
62.                    <args>
63.                        <arg>-Xjsr305=strict</arg>
64.                    </args>
```

```
65.                    <compilerPlugins>
66.                        <plugin>spring</plugin>
67.                        <plugin>jpa</plugin>
68.                    </compilerPlugins>
69.                </configuration>
70.                <dependencies>
71.                    <dependency>
72.                        <groupId>org.jetbrains.kotlin</groupId>
73.                        <artifactId>kotlin-maven-allopen</artifactId>
74.                        <version>${kotlin.version}</version>
75.                    </dependency>
76.                    <dependency>
77.                        <groupId>org.jetbrains.kotlin</groupId>
78.                        <artifactId>kotlin-maven-noarg</artifactId>
79.                        <version>${kotlin.version}</version>
80.                    </dependency>
81.                </dependencies>
82.            </plugin>
83.        </plugins>
84.    </build>
85. </project>
```

bootstrap.yml 檔案的內容如下所示，其中指定了設定檔的格式、目錄等資訊：

```
1.  spring:
2.    application:
3.      name: consul-config        # 應用名
4.    cloud:
5.      consul:
6.        host: 127.0.0.1          # Consul 設定中心位址
7.        port: 8500              # Consul 設定中心通訊埠
8.        config:
9.          prefix: config        # 設定檔字首
10.         enabled: true         # 是否生效
```

```
11.        format: yaml         # 設定檔格式
12.        data-key: data       # 設定項目字首
13.   profiles:
14.     active: dev             # 使用 dev 結尾的設定檔
15. server:
16.   port: 8065               # 應用通訊埠編號
```

ConsulConfigApplication.kt 定義了一個啟動類別：

```
1.  @SpringBootApplication
2.  class ConsulConfigApplication
3.  // 啟動函數
4.  fun main(args: Array<String>) {
5.      runApplication<ConsulConfigApplication>(*args)
6.  }
```

ConsulConfig.kt 定義了 Consul 中的設定項目。使用 @ConfigurationProperties
註釋將 ConsulConfig 定義的值和 Consul 中的設定對應起來：

```
1.  // 設定實體類別，對應 Consul 定義的設定
2.  @Component
3.  @ConfigurationProperties(prefix = "data")
4.  class ConsulConfig {
5.      var env: String? = null
6.      var user: User? = null
7.  }
8.  // User 實體類別
9.  class User{
10.     var username: String? = null
11.     var password: String? = null
12.     override fun toString(): String {
13.         return "${username},${password}"
14.     }
15. }
```

ConsulController.kt 定義了一個測試介面：

```
1.  @RestController
2.  class ConsulController {
3.      @Autowired
4.      lateinit var consulConfig: ConsulConfig
5.      // 測試介面，從 Consul 讀取設定
6.      @GetMapping("config")
7.      fun getConsulConfig(): String {
8.          return "${consulConfig.env}, ${consulConfig.user}"
9.      }
10. }
```

存取 "config" 可以從 Consul 取得設定值。在 Consul 中修改設定值，再次存取 "config" 介面，可以取得最新的設定。

6.5 小結

本章介紹了 Kotlin 整合 Spring Cloud Config、Apollo、Nacos、Consul 這四種設定中心的相關知識。對微服務系統來說，一個系統可以劃分為很多個微服務，每個微服務都有自己的設定。設定中心可幫助高效管理微服務的設定，設定的熱載入機制使得不需要重新啟動應用，設定就可生效。

Kotlin 應用於微服務閘道

閘道是微服務系統對外提供服務的視窗，透過閘道，可將請求分發到不同的子系統。本章將介紹 Zuul 和 Spring Cloud Gateway，並透過範例介紹 Kotlin 開發微服務閘道的方法。

7.1 Kotlin 整合 Zuul

Zuul 是 Netflix 公司開放原始碼的 API Gateway 元件，是所有從裝置和 Web 網站到 Netflix 串流媒體應用程式後端請求的前門。本節將介紹 Zuul 的架構和原理，以及使用 Kotlin 整合 Zuul 進行微服務閘道開發的相關知識。

7.1.1 Zuul 介紹

作為一個邊緣服務應用程式，Zuul 支援動態路由、監視，並為請求提供彈性和安全性。它還可以根據需要將請求路由到多個 Amazon 自動伸縮群組。Zuul 使用了一系列不同類型的篩檢程式，使我們能夠快速靈活地將功能應用到邊緣服務中。這些篩檢程式可幫助我們執行以下功能。

- 身份驗證和安全性：滿足每個資源的身份驗證需求並拒絕不滿足這些需求的請求。
- 洞察和監控：在邊緣追蹤有意義的資料和統計資料，以便提供準確的生產視圖。
- 動態路由：將外部的請求轉發到實際的微服務實例上，這是實現外部存取統一入口的基礎。
- 壓力測試：逐步增加叢集的流量，以評估效能。
- 減少負載：為每種類型的請求分配容量，並刪除超過限制的請求。
- 靜態回應處理：直接在邊緣建置一些回應，而非將它們轉發到內部叢集。
- 多區域彈性：跨 AWS 區域路由請求，以使 ELB 使用多樣化。

Zuul 目前有兩個版本。Zuul1 設計比較簡單，本質上就是一個同步 Servlet，採用多執行緒阻塞模型，同步阻塞模式的程式設計模型比較簡單，開發偵錯運行維護也比較簡單。但是執行緒上下文切換負擔大，連接數量受限制，延遲阻塞會耗盡執行緒資源。Zuul2 使用 Netty 實現非同步非阻塞程式設計模型，執行緒負擔小，連接數量易於擴充，但是程式設計模型複雜，開發偵錯運行維護複雜。本書使用 Zuul1 版本介紹。

Zuul1 篩檢程式的原理如圖 7.1 所示，篩檢程式分為四種：pre、post、routing 及 error。

Zuul2 基於 Netty，它首先執行預篩檢程式（入站篩檢程式），然後使用 Netty 客戶端代理請求，最後在執行後篩檢程式（出站篩檢程式）後傳回回應。

Zuul 和 Eureka 進行整合，將 Zuul 本身註冊為 Eureka 服務治理下的應用，Eureka 註冊的其他應用可以透過 Zuul 跳躍後存取。Spring Cloud 對 Zuul 進行了整合與增強，Zuul 預設使用的 HTTP 用戶端是 Apache HTTPClient，也可以使用 RESTClient 或 OkHttp3 中的 OkHttpClient。Zuul

預設和 Ribbon 結合實現負載平衡的功能。Zuul 的路由設定資訊可以放在 Spring Cloud Config 中，實現動態調整。

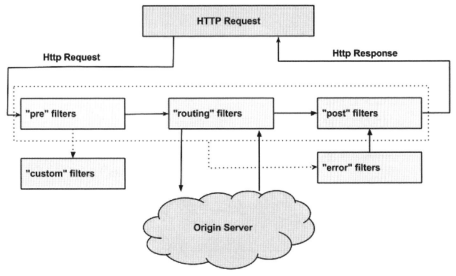

圖 7.1　Zuul1 篩檢程式的原理

7.1.2　Kotlin 整合 Zuul

新增 Maven 子專案 "chapter07-zuul"，這是一個以 Zuul 為基礎的閘道微服務。pom 檔案如下：

```xml
1.  <?xml version="1.0" encoding="UTF-8"?>
2.  <project xmlns="http://maven.apache.org/POM/4.0.0"
3.          xmlns:xsi="http://www.w3.org/2001/XMLSchema-instance"
4.          xsi:schemaLocation="http://maven.apache.org/POM/4.0.0 http://
    maven.apache.org/xsd/maven-4.0.0.xsd">
5.      <parent>
6.          <artifactId>kotlinspringboot</artifactId>
7.          <groupId>io.kang.kotlinspringboot</groupId>
8.          <version>0.0.1-SNAPSHOT</version>
9.      </parent>
```

```
10.    <modelVersion>4.0.0</modelVersion>
11.    <!-- 子專案名 -->
12.    <artifactId>chapter07-zuul</artifactId>
13.    <dependencies>
14.        <!-- Spring Cloud Zuul 相依套件 -->
15.        <dependency>
16.            <groupId>org.springframework.cloud</groupId>
17.            <artifactId>spring-cloud-starter-netflix-zuul</artifactId>
18.            <version>2.2.1.RELEASE</version>
19.        </dependency>
20.        <!-- Spring Cloud Config 相依套件 -->
21.        <dependency>
22.            <groupId>org.springframework.cloud</groupId>
23.            <artifactId>spring-cloud-starter-config</artifactId>
24.            <version>2.2.1.RELEASE</version>
25.        </dependency>
26.        <!-- Spring Cloud Eureka 用戶端相依套件 -->
27.        <dependency>
28.            <groupId>org.springframework.cloud</groupId>
29.            <artifactId>spring-cloud-starter-netflix-eureka-client
    </artifactId>
30.            <version>2.2.1.RELEASE</version>
31.        </dependency>
32.        <!-- Spring Boot Web 相依套件 -->
33.        <dependency>
34.            <groupId>org.springframework.boot</groupId>
35.            <artifactId>spring-boot-starter-web</artifactId>
36.            <version>2.2.1.RELEASE</version>
37.        </dependency>
38.        <!-- Spring Boot Actuator 相依套件 -->
39.        <dependency>
40.            <groupId>org.springframework.boot</groupId>
41.            <artifactId>spring-boot-starter-actuator</artifactId>
42.            <version>2.2.1.RELEASE</version>
43.        </dependency>
44.        <dependency>
```

```
45.          <groupId>com.fasterxml.jackson.module</groupId>
46.          <artifactId>jackson-module-kotlin</artifactId>
47.      </dependency>
48.      <dependency>
49.          <groupId>org.jetbrains.kotlin</groupId>
50.          <artifactId>kotlin-reflect</artifactId>
51.      </dependency>
52.      <dependency>
53.          <groupId>org.jetbrains.kotlin</groupId>
54.          <artifactId>kotlin-stdlib-jdk8</artifactId>
55.      </dependency>
56.      <dependency>
57.          <groupId>org.jetbrains.kotlinx</groupId>
58.          <artifactId>kotlinx-coroutines-core</artifactId>
59.          <version>1.3.2</version>
60.      </dependency>
61.   </dependencies>
62.   <build>
63.      <sourceDirectory>${project.basedir}/src/main/kotlin</sourceDirectory>
64.      <testSourceDirectory>${project.basedir}/src/test/kotlin
   </testSourceDirectory>
65.      <plugins>
66.          <plugin>
67.              <groupId>org.springframework.boot</groupId>
68.              <artifactId>spring-boot-maven-plugin</artifactId>
69.          </plugin>
70.          <plugin>
71.              <groupId>org.jetbrains.kotlin</groupId>
72.              <artifactId>kotlin-maven-plugin</artifactId>
73.              <configuration>
74.                  <args>
75.                      <arg>-Xjsr305=strict</arg>
76.                  </args>
77.                  <compilerPlugins>
78.                      <plugin>spring</plugin>
79.                      <plugin>jpa</plugin>
```

```
80.                    </compilerPlugins>
81.                </configuration>
82.                <dependencies>
83.                    <dependency>
84.                        <groupId>org.jetbrains.kotlin</groupId>
85.                        <artifactId>kotlin-maven-allopen</artifactId>
86.                        <version>${kotlin.version}</version>
87.                    </dependency>
88.                    <dependency>
89.                        <groupId>org.jetbrains.kotlin</groupId>
90.                        <artifactId>kotlin-maven-noarg</artifactId>
91.                        <version>${kotlin.version}</version>
92.                    </dependency>
93.                </dependencies>
94.            </plugin>
95.        </plugins>
96.    </build>
97. </project>
```

application.yml 檔案的內容如下所示，其中定義了路由資訊，介面 "/provide/*" 的呼叫被發送給 provider-server 這個微服務，對微服務 consumer-feign 不做代理：

```
1.  eureka:
2.    client:
3.      service-url:
4.        defaultZone: http://localhost:8761/eureka/    # Eureka 註冊中心位址
5.  server:
6.    port: 8070                                        # 應用通訊埠編號
7.  zuul:
8.    routes:
9.      provide-service-url:
10.       path: /provide/*                              # 請求位址
11.       service-id: provider-server                   # 被路由的服務 id
12.   ignored-services: consumer-feign                  # 不對該服務進行路由代理
```

```
13. ribbon:
14.   eureka:
15.     enabled: true                    # 使用 Ribbon 進行負載平衡
16. spring:
17.   application:
18.     name: zuul-app                   # 應用名稱
```

ZuulApplication.kt 是啟動類別，服務啟動後會註冊到 Eureka：

```
1.  // 開啟服務註冊註釋，Zuul 代理註釋
2.  @EnableDiscoveryClient
3.  @EnableZuulProxy
4.  @SpringBootApplication
5.  class ZuulApplication
6.  // 啟動函數
7.  fun main(args: Array<String>) {
8.      runApplication<ZuulApplication>(*args)
9.  }
```

首先啟動 Eureka 註冊中心——chapter05-eureka，然後啟動兩個服務——chapter05- eureka-provider、chapter05-eureka-consumer，它們在 Eureka 中的 service-id 是 provider-server、consumer-feign，再啟動閘道服務 chapter07-zuul。

呼叫 "provide/provide" 會轉發到 provider-server 的 "provide" 介面。如果 provider-server 有多個實例，透過設定 ribbon.eureka.enabled=true，Ribbon 會根據發現機制來取得設定服務名稱對應的實例清單，進行負載平衡。

Zuul 在註冊到 Eureka 服務中心之後，它會為 Eureka 中的每個服務都建立一個預設的路由規則，預設規則的 path 會使用 service-id 設定的服務名稱作為請求字首。在 application.yml 中將 ignored-services 註釋起來，保留 consumer-feign 這個微服務。呼叫 "consumer-feign/ feignProvide" 可以將請求轉發到 consumer-feign 的 "feignProvide" 介面。

Zuul 路由的 path 屬性通常需要萬用字元，萬用字元比對規則如下。

- ?：比對單一字元，如 "/feign/?"。
- *：比對任意數量字元，但不支援多級目錄，如 "/feign/*"。
- **：比對任意數量字元，支援多級目錄，如 "/feign/**"。

AccessFilter.kt 定義了一個篩檢程式，驗證請求是否包含 "accessToken" 參數。如果沒有 "accessToken"，傳回 401 碼值，如圖 7.2 所示。

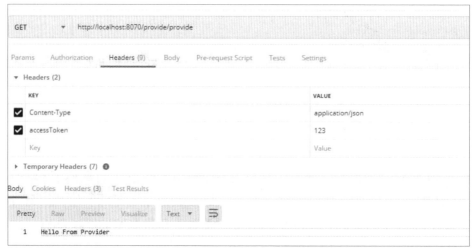

圖 7.2 /provide/provide 介面呼叫結果

```
1.   @Component
2.   class AccessFilter: ZuulFilter() {
3.       // 是否過濾
4.       override fun shouldFilter(): Boolean {
5.           return true
6.       }
7.       // 篩檢程式類型
8.       override fun filterType(): String {
9.           return "pre"
10.      }
11.      // 攔截 header 的 accessToken，不能為空
```

```
12.    override fun run(): Any? {
13.        val ctx = RequestContext.getCurrentContext()
14.        val request = ctx.request
15.        println(" 進入存取篩檢程式，存取的 url:${request.requestURL}，
   存取的方法：${request.method}")
16.        val accessToken = request.getHeader("accessToken")
17.        if(accessToken == null || accessToken.isEmpty()) {
18.            ctx.setSendZuulResponse(false)
19.            ctx.responseStatusCode = 401
20.            return null
21.        }
22.        return null
23.    }
24.    // 過濾的優先順序
25.    override fun filterOrder(): Int {
26.        return 0
27.    }
28. }
```

RateLimiterFilter.kt 定義了一個篩檢程式，驗證介面的存取頻率。介面存取
頻率過高，會傳回異常，提示呼叫過多。用 postman 對 "provide/provide"
進行平行處理呼叫，平行處理數為 5，結果如圖 7.3 所示。

圖 7.3 provide/provide 平行處理測試結果

```
1.  @Component
2.  class RateLimiterFilter: ZuulFilter() {
3.      // 初始化限流器
4.      val rateLimiter = RateLimiter.create(2.0)
5.      // 篩檢程式類型
6.      override fun filterType(): String {
7.          return PRE_TYPE
8.      }
9.      // 篩檢程式優先順序
10.     override fun filterOrder(): Int {
11.         return 1
12.     }
13.     // 對請求進行限流
14.     override fun run(): Any? {
15.         val ctx = RequestContext.getCurrentContext()
16.         if(!rateLimiter.tryAcquire()) {
17.             ctx.setSendZuulResponse(false)
18.             ctx.responseStatusCode = HttpStatus.TOO_MANY_REQUESTS.value()
19.         }
20.         return null
21.     }
22.     // 只對 /provide/limit 介面執行該篩檢程式
23.     override fun shouldFilter(): Boolean {
24.         val ctx = RequestContext.getCurrentContext()
25.         val request = ctx.request
26.         if("/provide/limit" == request.requestURI){
27.             return true
28.         }
29.         return false
30.     }
31. }
```

ApiFallbackProvider.kt 定義了一個熔斷器。如果其中一個服務「掛」掉了，那麼請求就會進行漫長的逾時等待，最後會傳回失敗，這甚至會影響整個服務鏈。熔斷器可以及時處理「掛」掉的服務，及時給使用者發送回應資訊：

```kotlin
1.  @Component
2.  class ApiFallbackProvider: FallbackProvider {
3.      // 對所有介面執行熔斷處理
4.      override fun getRoute(): String {
5.          return "*"
6.      }
7.      // 熔斷處理函數
8.      override fun fallbackResponse(route: String?, cause: Throwable?):
    ClientHttpResponse {
9.          var message = ""
10.         if (cause is HystrixTimeoutException) {
11.             message = "Timeout"
12.         } else {
13.             message = "Service exception"
14.         }
15.         return fallbackResponse(message)
16.     }
17.     // 封裝 ClienthttpResponse
18.     fun fallbackResponse(message: String): ClientHttpResponse {
19.         return object : ClientHttpResponse {
20.             // 傳回狀態碼
21.             @Throws(IOException::class)
22.             override fun getStatusCode(): HttpStatus {
23.                 return HttpStatus.OK
24.             }
25.             // 傳回狀態碼
26.             @Throws(IOException::class)
27.             override fun getRawStatusCode(): Int {
28.                 return 200
29.             }
30.             // 傳回狀態描述
31.             @Throws(IOException::class)
32.             override fun getStatusText(): String {
33.                 return "OK"
34.             }
35.             override fun close() {
```

```
36.              }
37.              // 傳回的 body
38.              @Throws(IOException::class)
39.              override fun getBody(): InputStream {
40.                  val bodyText = String.format("{\"code\": 999,\"message\":
     \"Service unavailable:%s\"}", message)
41.                  return ByteArrayInputStream(bodyText.toByteArray())
42.              }
43.              // HTTP 請求標頭
44.              override fun getHeaders(): HttpHeaders {
45.                  val headers = HttpHeaders()
46.                  headers.setContentType(MediaType.APPLICATION_JSON)
47.                  return headers
48.              }
49.          }
50.      }
51. }
```

將服務 "provider-server" 停掉，再次存取介面 "provide/provide" 會傳回預設值，如圖 7.4 所示。

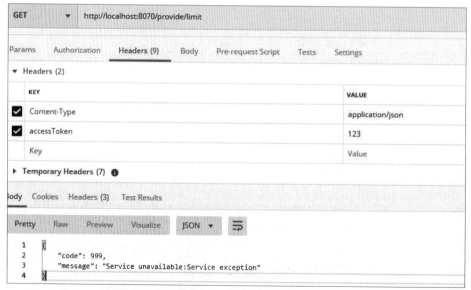

圖 7.4 停掉 "provider-server"，provide/provide 介面的呼叫結果

為了實現動態路由，可以將 Zuul 的路由設定放在設定中心，如 Spring Cloud Config。在 Spring Cloud Config 中新增一個檔案：zuulapp-dev.yml，上傳到 Git 倉庫：

```
1.  zuul:
2.    routes:
3.      provide-service-url:
4.        path: /provide/*  # 請求位址
5.        service-id: provider-server      # Zuul 代理的服務 id
6.    ignored-services: consumer-feign     # Zuul 隱藏的服務
```

chapter07-zuul 子專案的 application.yml 檔案調整如下：

```
1.  eureka:
2.    client:
3.      service-url:
4.        defaultZone: http://localhost:8761/eureka/     # Eureka 註冊中心位址
5.  server:
6.    port: 8070                       # 服務通訊埠編號
7.  ribbon:
8.    eureka:
9.      enabled: true                  # 使用 Ribbon 進行負載平衡
10. management:
11.   endpoints:
12.     web:
13.       exposure:
14.         include: refresh,health,info  # 對外曝露的監控介面
```

新增一個設定檔 bootstrap.yml：

```
1.  spring:
2.    cloud:
3.      config:
4.        label: master               # 設定檔所在分支
5.        profile: dev                 # dev 結尾的設定檔
6.        uri: http://localhost:8060   # 註冊中心位址
```

```
7.    application:
8.      name: zuulapp                    # 應用名
```

ZuulProperty.kt 定義了 Zuul 的相關設定：

```
1.    // 設定實體類別，對應 zuulapp-dev.yml 中定義的屬性
2.    @Configuration
3.    class ZuulProperty {
4.        @Bean
5.        @ConfigurationProperties("zuul")
6.        @RefreshScope
7.        @Primary
8.        fun zuulProperties(): ZuulProperties {
9.            return ZuulProperties()
10.       }
11.   }
```

依次啟動 chapter05-eureka、chapter05-eureka-provider、chapter05-eureka-consumer、chapter06-springcloud-config 和 chapter07-zuul，然後存取 "consumer-feign/feignProvide" 介面。由於 zuulconfig-dev.yml 設定中隱藏了 "consumer-feign" 這個服務，所以呼叫不能轉發到 "feignProvide" 介面。修改 zuulconfig-dev.yml，註釋起來 "ignored-services" 這個設定項目，並提交到遠端倉庫。然後在 chapter07-zuul 專案中呼叫 "actuator/refresh" 更新設定，再次存取 "consumer-feign/feignProvide" 介面，可以把服務轉發到 "feignProvide" 介面。

7.2 Kotlin 整合 Spring Cloud Gateway

Spring Cloud Gateway 是 Spring 官方基於 Spring 5.0、Spring Boot 2.0 和 Project Reactor 等技術開發的閘道，旨在為微服務架構提供一種簡單有效、統一的 API 路由管理方式，並統一存取介面。本節介紹 Kotlin 整合 Spring Cloud Gateway 進行微服務閘道開發的相關知識。

7.2.1 Spring Cloud Gateway 介紹

Spring Cloud Gateway 作為 Spring Cloud 生態系統中的閘道，目標是替代 Netflix Zuul，其不僅提供了統一的路由方式，並且以 Filter 鏈為基礎的方式提供了閘道應具備的基本功能，舉例來說，安全、監控 / 埋點和限流等。它是以 Nttey 為基礎的響應式開發模式，使用的 Web 架構是 WebFlux。WebFlux 是一個典型的非阻塞的非同步的架構，它的核心是以 Reactor 為基礎的相關 API 實現的，具有非阻塞、函數式程式設計特點。

Spring Cloud Gateway 的工作流程如圖 7.5 所示。

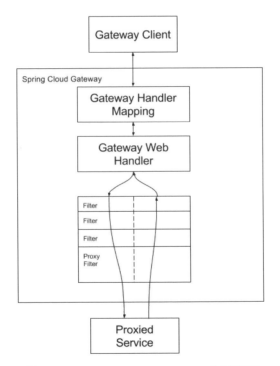

圖 7.5 Spring Cloud Gateway 工作流程圖

用戶端向 Spring Cloud Gateway 發出請求，然後在 Gateway Handler Mapping 中找到與請求相符合的路由，將其發送到 Gateway Web Handler。Handler

再透過指定的篩檢程式鏈將請求發送到實際的服務執產業務邏輯，然後傳回。篩檢程式之間用虛線分開是因為篩檢程式可能會在發送代理請求之前或之後執產業務邏輯。

路由（Route）是 Gateway 的基本建置模組。它由 ID、目標 URI、斷言集合和篩檢程式集合組成。如果聚合斷言結果為真，則比對到該路由。

斷言（Predicate）是一個 Java 8 Function Predicate。輸入類型是 Spring Framework ServerWebExchange，允許開發人員比對來自 HTTP 請求的任何內容，舉例來説，Header 或參數。

篩檢程式（Filter）是使用特定工廠建置的 Spring Framework Gateway Filter 實例，可以在傳回請求之前或之後修改請求和回應的內容。

Spring Cloud Gateway 整合 Hystrix 熔斷器、整合 Spring Cloud Discovery Client、整合開發人員易於撰寫的 Predicates 和 Filters，提供了以 Redis 為基礎的 Ratelimiter 實現，並使用權杖桶演算法限流，支援路徑重新定義，可以基於 Redis 使資料庫實現動態路由。

7.2.2 Kotlin 整合 Spring Cloud Gateway

新增一個 Maven 子專案：chapter07-gateway，這是一個基於 Spring Cloud Gateway 架設的閘道服務，pom 檔案如下：

```
1.  <?xml version="1.0" encoding="UTF-8"?>
2.  <project xmlns="http://maven.apache.org/POM/4.0.0"
3.          xmlns:xsi="http://www.w3.org/2001/XMLSchema-instance"
4.          xsi:schemaLocation="http://maven.apache.org/POM/4.0.0 http://
    maven.apache.org/xsd/maven-4.0.0.xsd">
5.      <parent>
6.          <artifactId>kotlinspringboot</artifactId>
7.          <groupId>io.kang.kotlinspringboot</groupId>
```

```xml
8.          <version>0.0.1-SNAPSHOT</version>
9.      </parent>
10.     <modelVersion>4.0.0</modelVersion>
11.     <!-- 子專案名 -->
12.     <artifactId>chapter07-gateway</artifactId>
13.     <dependencies>
14.         <!-- Spring Cloud Gateway 相依套件 -->
15.         <dependency>
16.             <groupId>org.springframework.cloud</groupId>
17.             <artifactId>spring-cloud-starter-gateway</artifactId>
18.             <version>2.2.1.RELEASE</version>
19.         </dependency>
20.         <!-- Spring Cloud Eureka 用戶端相依套件 -->
21.         <dependency>
22.             <groupId>org.springframework.cloud</groupId>
23.             <artifactId>spring-cloud-starter-netflix-eureka-client
    </artifactId>
24.             <version>2.2.1.RELEASE</version>
25.         </dependency>
26.         <!-- Spring Boot Redis 相依套件 -->
27.         <dependency>
28.             <groupId>org.springframework.boot</groupId>
29.             <artifactId>spring-boot-starter-data-redis-reactive
    </artifactId>
30.             <version>2.2.4.RELEASE</version>
31.         </dependency>
32.         <!-- Spring Cloud Hystrix 相依套件 -->
33.         <dependency>
34.             <groupId>org.springframework.cloud</groupId>
35.             <artifactId>spring-cloud-starter-netflix-hystrix</artifactId>
36.             <version>2.2.1.RELEASE</version>
37.         </dependency>
38.         <!-- Spring Boot Webflux 相依套件 -->
39.         <dependency>
40.             <groupId>org.springframework.boot</groupId>
```

```
41.              <artifactId>spring-boot-starter-webflux</artifactId>
42.              <version>2.2.1.RELEASE</version>
43.          </dependency>
44.          <!-- Fastjson 相依套件 -->
45.          <dependency>
46.              <groupId>com.alibaba</groupId>
47.              <artifactId>fastjson</artifactId>
48.              <version>1.2.62</version>
49.          </dependency>
50.          <!-- Spring Boot Actuator 相依套件 -->
51.          <dependency>
52.              <groupId>org.springframework.boot</groupId>
53.              <artifactId>spring-boot-starter-actuator</artifactId>
54.              <version>2.2.1.RELEASE</version>
55.          </dependency>
56.          <dependency>
57.              <groupId>com.fasterxml.jackson.module</groupId>
58.              <artifactId>jackson-module-kotlin</artifactId>
59.          </dependency>
60.          <dependency>
61.              <groupId>org.jetbrains.kotlin</groupId>
62.              <artifactId>kotlin-reflect</artifactId>
63.          </dependency>
64.          <dependency>
65.              <groupId>org.jetbrains.kotlin</groupId>
66.              <artifactId>kotlin-stdlib-jdk8</artifactId>
67.          </dependency>
68.          <dependency>
69.              <groupId>org.jetbrains.kotlinx</groupId>
70.              <artifactId>kotlinx-coroutines-core</artifactId>
71.              <version>1.3.2</version>
72.          </dependency>
73.          <!-- Spring Boot Test 相依套件 -->
74.          <dependency>
75.              <groupId>org.springframework.boot</groupId>
```

```
76.              <artifactId>spring-boot-starter-test</artifactId>
77.              <version>2.2.1.RELEASE</version>
78.          </dependency>
79.      </dependencies>
80.      <build>
81.          <sourceDirectory>${project.basedir}/src/main/kotlin
     </sourceDirectory>
82.          <testSourceDirectory>${project.basedir}/src/test/kotlin
     </testSourceDirectory>
83.          <plugins>
84.              <plugin>
85.                  <groupId>org.springframework.boot</groupId>
86.                  <artifactId>spring-boot-maven-plugin</artifactId>
87.              </plugin>
88.              <plugin>
89.                  <groupId>org.jetbrains.kotlin</groupId>
90.                  <artifactId>kotlin-maven-plugin</artifactId>
91.                  <configuration>
92.                      <args>
93.                          <arg>-Xjsr305=strict</arg>
94.                      </args>
95.                      <compilerPlugins>
96.                          <plugin>spring</plugin>
97.                          <plugin>jpa</plugin>
98.                      </compilerPlugins>
99.                  </configuration>
100.                 <dependencies>
101.                     <dependency>
102.                         <groupId>org.jetbrains.kotlin</groupId>
103.                         <artifactId>kotlin-maven-allopen</artifactId>
104.                         <version>${kotlin.version}</version>
105.                     </dependency>
106.                     <dependency>
107.                         <groupId>org.jetbrains.kotlin</groupId>
108.                         <artifactId>kotlin-maven-noarg</artifactId>
```

```
109.                        <version>${kotlin.version}</version>
110.                    </dependency>
111.                </dependencies>
112.            </plugin>
113.        </plugins>
114.    </build>
115. </project>
```

application.yml 檔案中的定義如下：定義了對 provide-server 服務的路由對映，/provide/** 的路徑會被轉發到 provide-server 的 "/**" 介面；StripPrefix=1 表示忽略第一個路徑，即 /provide。定義了一個服務不可用，熔斷的預設介面為 "/fallback"。定義了一個限流器，其根據 apiKeyResolver 定義的 Bean 的名字，對對應的 bean 的呼叫進行限流：

```
1.  server:
2.    port: 8071              # 應用通訊埠編號
3.  spring:
4.    application:
5.      name: gateway-app    # 應用名
6.    cloud:
7.      gateway:
8.        discovery:
9.          locator:
10.            enabled: true    # 是否和服務註冊與發現元件結合，設定為 true 後
                                # 可以直接使用應用名稱呼叫服務
11.        routes:
12.          - id: provide-server          # 被路由代理的服務 id
13.            uri: lb://provider-server    # 被路由代理的 uri
14.            predicates:
15.              - Path=/provide/**          # 被代理的 url
16.            filters:
17.              ## 截取路徑位元數
18.              - StripPrefix=1
19.              - name: Hystrix             # Hystrix 設定
```

```
20.          args:
21.            name: fallBackBean          # 熔斷器名稱
22.            fallbackUri: forward:/fallback  # 提供熔斷的介面
23.          - name: RequestRateLimiter      # 限流器設定
24.            args:
25.              ### 限流篩檢程式的 Bean 名稱
26.              key-resolver: '#{@apiKeyResolver}'
27.              ### 希望允許使用者每秒處理多少個請求
28.              redis-rate-limiter.replenishRate: 1
29.              ### 使用者允許在 1 秒內完成的最大請求數
30.              redis-rate-limiter.burstCapacity: 3
31.    redis:
32.      host: localhost      # Redis 連接 host
33.      prot: 6379           # Redis 連接通訊埠
34.      password: 123456     # Redis 連接密碼
35. eureka:
36.    client:
37.      service-url:
38.        defaultZone: http://localhost:8761/eureka/    # Eureka 註冊中心位址
```

GateWayApplication.kt 定義了一個啟動類別，服務啟動後會註冊到 Eureka：

```
1.  @SpringBootApplication
2.  @EnableEurekaClient
3.  class GateWayApplication
4.  // 啟動函數
5.  fun main(args: Array<String>) {
6.      runApplication<GateWayApplication>(*args)
7.  }
```

順序啟動 chapter05-eureka、chapter05-eureka-provider、chapter07-gateway，呼叫 chapter07-gateway 的 "/provide/provide" 可以將請求轉發到 chapter05-eureka-provider 的 "provide" 介面。

FallbackController.kt 定義了服務不可用時預設的介面：

```
1.  class FallbackController {
2.      // 提供熔斷服務的介面
3.      @GetMapping("/fallback")
4.      fun fallback(): String {
5.          return "I'm Spring Cloud Gateway fallback."
6.      }
7.  }
```

關閉 chapter05-eureka-provider 服務，再次呼叫 "/provide/provide"，會轉到 "/fallback" 介面，傳回 "I'm Spring Cloud Gateway fallback."，如圖 7.6 所示。

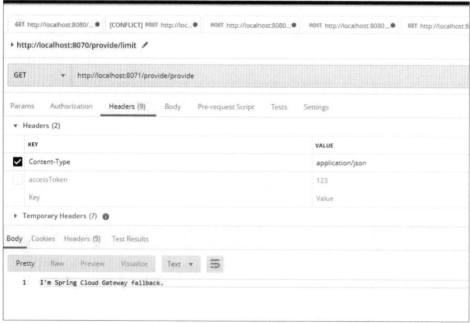

圖 7.6 /provide/provide 介面服務降級呼叫的結果

RequestRateLimiterConfig.kt 定義了幾種限流策略。ipAddressKeyResolver 根據呼叫伺服器的主機名稱限流，apiKeyResolver 根據呼叫的路徑進行限流，userKeyResolver 根據查詢參數 userId 進行限流，yml 檔案定義的限流策略是 apiKeyResolver：

```kotlin
1.  @Configuration
2.  class RequestRateLimiterConfig {
3.      // 根據主機名稱對請求進行限流
4.      @Bean("ipAddressKeyResolver")
5.      fun ipAddressKeyResolver(): KeyResolver {
6.          return KeyResolver {
7.              exchange -> Mono.just(exchange.request.remoteAddress?.
   hostName.orEmpty())
8.          }
9.      }
10.     // 根據呼叫路徑對請求進行限流
11.     @Bean("apiKeyResolver")
12.     @Primary
13.     fun apiKeyResolver(): KeyResolver {
14.         return KeyResolver {
15.             exchange -> Mono.just(exchange.request.path.value())
16.         }
17.     }
18.     // 根據請求參數的 userId 對請求進行限流
19.     @Bean("userKeyResolver")
20.     fun userKeyResolver(): KeyResolver {
21.         return KeyResolver {
22.             exchange -> Mono.just(exchange.request.queryParams.getFirst
   ("userId").orEmpty())
23.         }
24.     }
25. }
```

使用 postman 對介面 "/provide/provide" 做平行處理測試，平行處理數為
10，只有 4 次呼叫傳回正常，其他呼叫都被限制了，如圖 7.7 所示。

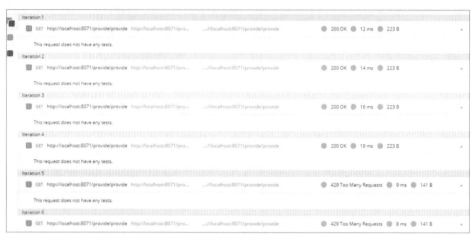

圖 7.7 provide/provide 介面平行處理限流測試結果

RedisRouteRepository.kt 將路由快取到 Redis，實現對路由設定的動態更
新。save 方法可以新增路由設定，getRouteDefinitions 方法傳回所有的路
由設定，delete 方法根據路由 id 刪除路由設定。程式如下：

```
1.    @Component
2.    class RedisRouteRepository: RouteDefinitionRepository {
3.        // 植入 redisTemplate
4.        @Autowired
5.        lateinit var redisTemplate: RedisTemplate<String, String>
6.        val routeKey = "route"
7.        // 儲存路由設定
8.        override fun save(route: Mono<RouteDefinition>): Mono<Void> {
9.            route
10.                   .subscribe { routeDefinition ->
11.                       println("${JSON.toJSONString(routeDefinition)}")
12.                       redisTemplate.opsForHash<String, String>().put(
13.           routeKey, routeDefinition.id,  JSON.toJSONString(routeDefinition))
```

```kotlin
14.                     }
15.             return Mono.empty<Void>()
16.     }
17.     // 取得快取中所有的路由設定
18.     override fun getRouteDefinitions(): Flux<RouteDefinition> {
19.         if (redisTemplate.hasKey(routeKey)) {
20.             // 從 Redis 中拉取路由
21.             val routeDefinitions = LinkedList<RouteDefinition>()
22.             redisTemplate
23.                     .opsForHash<String, String>().values(routeKey)
24.                     .stream()
25.                     .forEach { routeDefinition -> routeDefinitions.
    add(JSON.parseObject(routeDefinition, RouteDefinition::class.java) as
    RouteDefinition) }
26.             return Flux.fromIterable(routeDefinitions)
27.         } else {
28.             var routes = LinkedHashMap<String, String>()
29.             redisTemplate.opsForHash<String, String>().putAll(routeKey,
    routes)
30.             return Flux.fromIterable(LinkedList<RouteDefinition>())
31.         }
32.     }
33.     // 根據 routeId 從快取中刪除路由設定
34.     override fun delete(routeId: Mono<String>): Mono<Void> {
35.         routeId
36.                 .subscribe { routeId ->
37.                     if (redisTemplate.opsForHash<String, String>().
    hasKey(routeKey, routeId)) {
38.                         redisTemplate.opsForHash<String, String>().
    delete(routeKey, routeId)
39.                     }
40.                 }
41.         return Mono.empty<Void>()
42.     }
43. }
```

RouteController.kt 定義了增加、查詢、刪除路由的介面:

```
1.   @RestController
2.   @RequestMapping("/route")
3.   class RouteController {
4.       @Autowired
5.       lateinit var redisRouteRepository: RedisRouteRepository
6.       // 測試介面,增加路由
7.       @PostMapping("/add")
8.       fun add(@RequestBody routeDefinition: RouteDefinition): Mono<String> {
9.           redisRouteRepository.save(Mono.just(routeDefinition))
10.          return Mono.just("add ok")
11.      }
12.      // 測試介面,取得所有路由
13.      @GetMapping("/all")
14.      fun getAll(): Flux<RouteDefinition> {
15.          return redisRouteRepository.routeDefinitions
16.      }
17.      // 測試介面,刪除指定 id 路由
18.      @DeleteMapping("/{id}")
19.      fun delete(@PathVariable id: String): Mono<String> {
20.          redisRouteRepository.delete(Mono.just(id))
21.          return Mono.just("delete ok")
22.      }
23. }
```

啟動 chapter05-eureka-consumer 專案,使用 postman 呼叫 "/route/add" 介面
增加一個新的路由設定,如圖 7.8 所示。

增加了對 consumer-feign 這個服務的路由設定,呼叫 "/feignProvide/
feignProvide" 介面,可以將請求轉發到 consumer-feign 的 "feignProvide"
介面。呼叫 "/route/consumer-feign" 介面,可刪除該筆設定,再次呼叫 "/
feignProvide/feignProvide",呼叫不通。透過 Redis,可以動態更新快取設
定,實現路由設定熱載入。

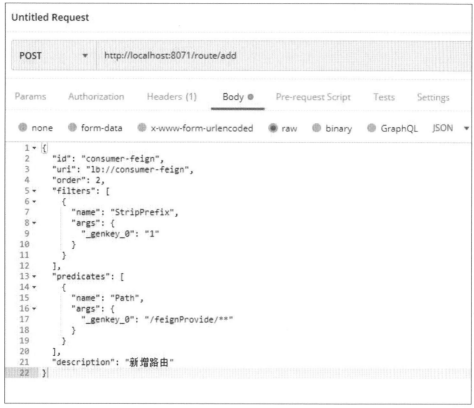

圖 7.8 介面 route/add 增加路由呼叫示意圖

7.3 小結

閘道隱藏內部細節，為呼叫者提供統一入口。閘道接收所有呼叫者請求，並透過路由機制轉發到服務實例。閘道是一組「篩檢程式」集合，可以實現一系列與核心業務無關的橫切面功能，如安全認證、限流熔斷、記錄檔監控等。

本節介紹了兩種閘道：Zuul 和 Spring Cloud Gateway。Spring Cloud 對它們都提供了很好的整合，使用起來很方便。Zuul1 基於 Servlet，Zuul2 基於 Netty，是非阻塞的，響應式的；Spring Cloud Gateway 基於 Reactor，是非同步非阻塞的。Zuul1 採用同步阻塞方式，效能較差，Spring Cloud Gateway 和 Zuul2 效能較好。

Chapter 08

Kotlin 應用於 Spring Cloud Alibaba

Spring Cloud Alibaba 是阿里巴巴集團開放原始碼的微服務開發的整合式解決方案。該專案包含開發分散式應用服務的必備元件,方便開發者透過 Spring Cloud 程式設計模型輕鬆使用這些元件來開發分散式應用服務。依靠 Spring Cloud Alibaba,只需增加一些註釋和少量設定,就可以將 Spring Cloud 應用連線阿里巴巴分散式應用解決方案,透過阿里巴巴中介軟體來迅速架設分散式應用系統。

Spring Cloud Alibaba 包含開放原始碼元件和商業化元件。開放原始碼元件包含:分散式設定管理、服務註冊與發現、服務限流降級、訊息中介軟體、Dubbo Spring Cloud 及分散式交易。商業化元件包含:商業版服務註冊與發現、商業版應用設定管理、物件儲存服務、分散式任務排程及通訊服務。

下面詳細介紹其中的一些元件。

服務限流降級: 預設支援 WebServlet、WebFlux、OpenFeign、Rest Template、Spring Cloud Gateway、Zuul、Dubbo 和 RocketMQ 等元件限流降級功能的連線,可以在執行時期透過主控台即時修改限流降級規則,還支援檢視限流降級指標監控。

服務註冊與發現：轉換 Spring Cloud 服務註冊與發現標準，預設整合了 Ribbon 的支援。

分散式設定管理：支援分散式系統中的外部化設定，設定更改時自動更新。

訊息驅動能力：基於 Spring Cloud Stream 為微服務應用建置訊息驅動能力。

分散式交易：使用 @GlobalTransactional 註釋，高效且對業務零侵入地解決分散式交易問題。

分散式任務排程：提供秒級、精準、高可靠、高可用的定時（基於 Cron 運算式）任務排程服務。同時提供分散式的任務執行模型，如網格任務。網格任務支援將海量子任務均勻分配到所有 Worker（schedulerx-client）上執行。

阿里雲簡訊服務：覆蓋全球的簡訊服務，人性化、高效、智慧的互連化通訊能力，幫助企業迅速架設客戶觸達通道。

Sentinel：把流量管理作為切入點，從流量控制、熔斷降級、系統負載保護等多個維度保護服務的穩定性。

Nacos：提供了一個易於建置雲原生應用的動態服務發現、設定管理和服務管理平台。

RocketMQ：一款開放原始碼的分散式訊息系統，基於高可用分散式叢集技術，提供低延遲時間、高可靠的訊息發佈與訂閱服務。

Dubbo：一款高性能的 Java RPC 架構。

Seata：一個易用的高性能微服務分散式交易解決方案。

Alibaba Cloud ACM：一款在分散式架構環境中對應用設定進行集中管理和發送的應用設定中心產品。

Alibaba Cloud OSS：阿里雲物件儲存服務（Object Storage Service，簡稱 OSS），提供巨量、安全、低成本、高可靠的雲端儲存服務。可以在任何應用、任何時間、任何地點儲存和存取任意類型的資料。

8.1 服務限流降級

Sentinel 誕生於 2012 年，其主要功能是控制介面流量。2013—2017 年，Sentinel 在阿里巴巴集團內部使用，2018 年，Sentinel 被開放原始碼。本節主要介紹 Kotlin 整合 Sentinel 實現服務限流、降級的方法。

8.1.1 Sentinel 介紹

資源是 Sentinel 的關鍵概念，它可以是 Java 應用程式中的任何內容，舉例來說，由應用程式提供的服務，或由應用程式呼叫其他應用提供的服務，甚至是一段程式。

規則是圍繞資源的即時狀態設定的，包含流量控制規則、熔斷降級規則、系統負載保護規則，所有規則都可以動態即時調整。

流量控制用於調整網路封包發送的資料，包含：資源的呼叫關係，舉例來說，資源的呼叫鏈路、資源和資源之間的關係；執行指標，舉例來說，QPS、執行緒池、系統負載等；控制的方式，舉例來說，直接限流、冷啟動、排隊。Sentinel 作為一個轉換器，可以根據需要把隨機的請求調整成合適的形狀，如圖 8.1 所示。

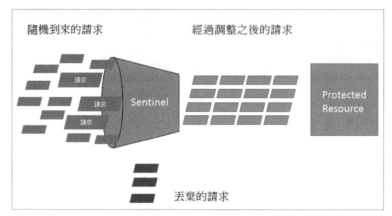

圖 8.1 Sentinel 功能示意圖

Sentinel 和 Hystrix 的原則一致，當呼叫鏈路中某個資源出現不穩定情況時，例如逾時、異常比例升高，對這個資源的呼叫進行限制，並讓請求快速失敗，避免影響到其他資源。在限制的方法上，Sentinel 和 Hystrix 採取了完全不同的方法。

Hystrix 透過執行緒池的方式對依賴（對應 Sentinel 中的資源）進行隔離。這樣做的好處是資源和資源之間做到了徹底隔離；缺點是增加了執行緒切換的成本，需要預先給各個資源進行執行緒池大小的分配。

Sentinel 透過平行處理執行緒數進行限制，限制資源平行處理執行緒的數量來減少不穩定資源對其他資源的影響。這樣不但可減少執行緒切換的損耗，也不需要預先分配執行緒池大小。當某個資源出現不穩定的情況時，舉例來說，回應時間變長，會造成執行緒數的逐步堆積。當執行緒數在特定資源上堆積到一定數量之後，對該資源的新請求就會被拒絕。堆積的執行緒完成任務後才會開始繼續接收請求。此外，Sentinel 透過回應時間對資源進行降級，透過回應時間來快速降級不穩定的資源。當依賴的資源出現回應時間過長時，所有對資源的存取都會被直接拒絕，直到過了指定的時間之後才會重新恢復。

Sentinel 同時對系統的維度提供保護，防止「雪崩」。當系統負載較高時，如果還持續讓請求進入，可能會導致系統當機、無法回應。在叢集環境下，網路負載平衡會把本應這台機器承載的流量轉發到其他機器上。如果這時那個「其他機器」也處於邊緣狀態，那麼增加的流量就會導致這台機器也當機，最後導致整個叢集不可用。

8.1.2 Kotlin 整合 Sentinel

新增一個 Maven 子專案：chapter08-sentinel，使用 Sentinel 對介面進行限流。pom 檔案如下：

```
1.  <?xml version="1.0" encoding="UTF-8"?>
2.  <project xmlns="http://maven.apache.org/POM/4.0.0"
3.          xmlns:xsi="http://www.w3.org/2001/XMLSchema-instance"
4.          xsi:schemaLocation="http://maven.apache.org/POM/4.0.0 http://
    maven.apache.org/xsd/maven-4.0.0.xsd">
5.      <parent>
6.          <artifactId>kotlinspringboot</artifactId>
7.          <groupId>io.kang.kotlinspringboot</groupId>
8.          <version>0.0.1-SNAPSHOT</version>
9.      </parent>
10.     <modelVersion>4.0.0</modelVersion>
11.     <!-- 子專案名 -->
12.     <artifactId>chapter08-sentinel</artifactId>
13.     <dependencies>
14.         <!-- Spring Cloud Sentinel 相依套件 -->
15.         <dependency>
16.             <groupId>com.alibaba.cloud</groupId>
17.             <artifactId>spring-cloud-starter-alibaba-sentinel</artifactId>
18.             <version>2.2.0.RELEASE</version>
19.         </dependency>
20.         <!-- Spring Boot Web 相依套件 -->
21.         <dependency>
22.             <groupId>org.springframework.boot</groupId>
```

```
23.          <artifactId>spring-boot-starter-web</artifactId>
24.          <version>2.2.1.RELEASE</version>
25.      </dependency>
26.      <!-- Sentinel Nacos 相依套件 -->
27.      <dependency>
28.          <groupId>com.alibaba.csp</groupId>
29.          <artifactId>sentinel-datasource-nacos</artifactId>
30.          <version>1.7.1</version>
31.      </dependency>
32.      <dependency>
33.          <groupId>com.fasterxml.jackson.module</groupId>
34.          <artifactId>jackson-module-kotlin</artifactId>
35.      </dependency>
36.      <dependency>
37.          <groupId>org.jetbrains.kotlin</groupId>
38.          <artifactId>kotlin-reflect</artifactId>
39.      </dependency>
40.      <dependency>
41.          <groupId>org.jetbrains.kotlin</groupId>
42.          <artifactId>kotlin-stdlib-jdk8</artifactId>
43.      </dependency>
44.      <dependency>
45.          <groupId>org.jetbrains.kotlinx</groupId>
46.          <artifactId>kotlinx-coroutines-core</artifactId>
47.          <version>1.3.2</version>
48.      </dependency>
49.  </dependencies>
50.  <build>
51.      <sourceDirectory>${project.basedir}/src/main/kotlin
</sourceDirectory>
52.      <testSourceDirectory>${project.basedir}/src/test/kotlin
</testSourceDirectory>
53.      <plugins>
54.          <plugin>
55.              <groupId>org.springframework.boot</groupId>
```

```
56.                     <artifactId>spring-boot-maven-plugin</artifactId>
57.                 </plugin>
58.                 <plugin>
59.                     <groupId>org.jetbrains.kotlin</groupId>
60.                     <artifactId>kotlin-maven-plugin</artifactId>
61.                     <configuration>
62.                         <args>
63.                             <arg>-Xjsr305=strict</arg>
64.                         </args>
65.                         <compilerPlugins>
66.                             <plugin>spring</plugin>
67.                             <plugin>jpa</plugin>
68.                         </compilerPlugins>
69.                     </configuration>
70.                     <dependencies>
71.                         <dependency>
72.                             <groupId>org.jetbrains.kotlin</groupId>
73.                             <artifactId>kotlin-maven-allopen</artifactId>
74.                             <version>${kotlin.version}</version>
75.                         </dependency>
76.                         <dependency>
77.                             <groupId>org.jetbrains.kotlin</groupId>
78.                             <artifactId>kotlin-maven-noarg</artifactId>
79.                             <version>${kotlin.version}</version>
80.                         </dependency>
81.                     </dependencies>
82.                 </plugin>
83.             </plugins>
84.         </build>
85. </project>
```

application.yml 檔案的內容如下：

```
1.  server:
2.    port: 8081                          # 應用通訊埠編號
3.  spring:
```

```
4.    application:
5.      name: sentinel-app      #應用名
6.    cloud:
7.      sentinel:
8.        datasource:
9.          ds1:
10.           file:
11.             file: classpath:flowrule.json #限流設定檔名
12.             data-type: json #檔案格式
13.             rule-type: flow #限流規則類型:以 JSON 格式傳回現有的限流規則
14.          ds2:
15.            nacos:
16.              server-addr: localhost:8848      # Nacos 設定中心位址
17.              data-id: ${spring.application.name}-rule   #Nacos 設定的 data-id
18.              group-id: DEFAULT_GROUP          # Nacos 設定的 group
19.              rule-type: flow #限流規則類型:以 JSON 格式傳回現有的限流規則
20.        transport:
21.          dashboard: localhost:8080            # 應用通訊埠編號
22.        eager: true                           # 應用通訊埠編號
```

Sentinel 的各種保護規則可以儲存在 JSON 檔案中,也可以儲存在 Nacos 等設定中心。application.yml 定義了兩種儲存規則的方式:flowrule.json 和 nacos。flowrule.json 檔案如下:

```
1.  [
2.    {
3.      "resource": "helloApi",
4.      "controlBehavior": 1,
5.      "count": 10,
6.      "grade": 1,
7.      "limitApp": "default",
8.      "strategy": 0
9.    },
10.   {
11.     "resource": "helloApi1",
12.     "controlBehavior": 2,
```

```
13.     "count": 10,
14.     "grade": 1,
15.     "limitApp": "default",
16.     "strategy": 0
17.   }
18. ]
```

resource 代表資源，是限流規則的作用物件；count 是限流設定值；grade
是限流閾數值型態，0 表示基於執行緒數，1 表示以 QPS；limitApp
是流量控制針對為基礎的呼叫來源，default 表示不區分呼叫來源；
controlBehavior 是流量控制方式，0 表示預設，1 表示冷啟動，2 表示勻速
排隊等待，3 表示慢啟動；strategy 表示根據什麼限流，0 表示根據資源本
身，1 表示連結其他資源，2 表示根據鏈路入口。

Nacos 設定的限流規則如圖 8.2 所示。

圖 8.2　Nacos 設定的限流規則

SentinelApplication.kt 定義了一個啟動類別，啟動一個 Spring Boot 應用：

```
1.  @SpringBootApplication
2.  class SentinelApplication {
3.      @Bean
4.      fun sentinelResourceAspect(): SentinelResourceAspect {
5.          return SentinelResourceAspect()
6.      }
7.  }
8.  // 啟動函數
9.  fun main(args: Array<String>) {
10.     runApplication<SentinelApplication>(*args)
11. }
```

SentinelController.kt 定義了三個介面，用於測試限流規則：

```
1.  @RestController
2.  class SentinelController {
3.      // 測試介面，被限流後預設呼叫 handleException 函數
4.      @GetMapping("/hello")
5.      @SentinelResource(value = "helloApi", blockHandler =
    "handleException", blockHandlerClass = [ExceptionUtil::class])
6.      fun hello(): String {
7.          return "hello sentinel"
8.      }
9.      // 測試介面，被限流後預設呼叫 exceptionHandler 函數
10.     @GetMapping("/hello1")
11.     @SentinelResource(value = "helloApi1", blockHandler = "exceptionHandler")
12.     fun hello1(): String {
13.         return "hello1 sentinel"
14.     }
15.     // 測試介面，從 Nacos 讀取限流規則，被限流後預設呼叫 exceptionHandler 函數
16.     @GetMapping("/hello2")
17.     @SentinelResource(value = "helloNacosApi", blockHandler =
    "exceptionHandler")
```

```
18.    fun helloNacos(): String {
19.        return "hello1 sentinel in nacos"
20.    }
21.    // 被限流後預設呼叫 exceptionHandler 函數
22.    fun exceptionHandler(s: Long, ex: BlockException): String {
23.        println(ex.printStackTrace())
24.        return "Oops, error occurred at $s"
25.    }
26. }
```

ExceptionUtil.kt 定義異常處理方法：

```
1.  // 異常處理方法
2.  class ExceptionUtil {
3.      fun handleException(ex: BlockException): String {
4.          return "Oops:${ex.javaClass.canonicalName}"
5.      }
6.  }
```

"/hello" 和 "/hello1" 兩個介面的限流規則儲存在 flowrule.json 檔案中，
"/hello2" 限流規則儲存在 Nacos 中。將 Nacos 的 controlBehavior 改為
0，使用 postman 呼叫 "/hello2" 介面 10 次，每次呼叫都正常傳回；將
controlBehavior 改為 1，呼叫 "/hello2" 介面 10 次，只有三次正常傳回，
其餘 7 次呼叫被限制；將 controlBehavior 改為 2，呼叫 "/hello2" 介面 10
次，介面勻速排隊，呼叫正常；將 controlBehavior 改為 3，呼叫 "/hello2"
介面 10 次，介面緩慢排隊，呼叫正常，如圖 8.3 所示。

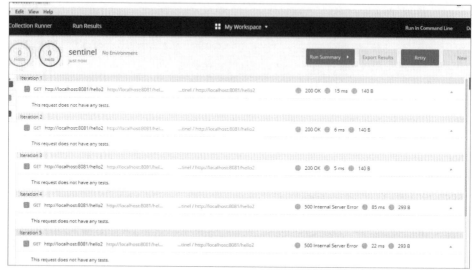

圖 8.3 介面 hello2 平行處理限流呼叫結果

8.2 訊息驅動

訊息驅動對於微服務系統很重要，有助系統間解耦。Spring Cloud Stream
是一個為微服務應用建置訊息驅動能力的架構。透過使用 Spring Cloud
Stream，可以有效降低訊息中介軟體的使用複雜度，讓系統開發人員可
以有更多的精力關注於核心業務邏輯的處理。本節介紹使用 Kotlin 整合
Spring Cloud Stream、RocketMQ 開發的相關知識。

8.2.1 訊息驅動介紹

Spring Cloud Stream 為一些訊息中介軟體提供了個性化的自動化設定功
能，並引用了發佈 - 訂閱、消費群組、分區這三個核心概念。RocketMQ
可以和 Spring Cloud Stream 整合。

應用程式透過 inputs、outputs 與 Spring Cloud Stream 中的 Binder 互動，
Spring Cloud Stream 中的 Binder 負責與訊息中介軟體互動。只需要弄清楚
如何與 Spring Cloud Stream 互動，就可以方便使用訊息驅動的方式，如圖
8.4 所示。

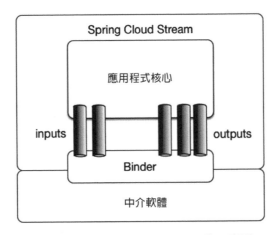

圖 8.4 Spring Cloud Stream 互動示意圖

透過使用 Spring Integration 連接訊息代理中介軟體就可以實現訊息事件驅
動。企業應用整合是一種應用之間資料和服務整合的技術，有以下整合風
格。

- 共用資料庫，兩個系統查詢同一個資料庫以取得要傳遞的資料。舉例來
  說，部署了兩個 Spring Boot 應用，它們的實體類別共用同一個表。
- 遠端程序呼叫，兩個系統都曝露另一個能呼叫的服務。舉例來說，EJB
  服務、SOAP 或 REST 服務。
- 訊息，兩個系統連接到一個公用的訊息系統，互相交換資料，並利用訊
  息呼叫行為。舉例來說，中心輻射式的（hub-and-spoke）JMS 架構。

Spring Cloud Stream 提供了一種應用和訊息中介軟體解耦的方式。Spring
Cloud Stream 由一個中介軟體中立的核心組成。inputs，相當於消費者

consumer，它從佇列中接收訊息；outputs，相當於生產者 producer，它從佇列中發送訊息。應用透過 inputs 和 outputs 通道與外界交流。通道透過指定中介軟體的 Binder 實現與外部代理連接。業務開發者不用關注實際的訊息中介軟體，只需關注 Binder 對應用程式提供的抽象概念以使用訊息中介軟體實現業務即可。

Binder 透過定義綁定器，並將其作為中間層，實現了應用程式與訊息中介軟體細節之間的隔離。透過向應用程式曝露統一的 Channel 通道，使應用程式不需要考慮各種不同的訊息中介軟體的實現細節。當需要升級訊息中介軟體，或更換其他訊息中介軟體產品時，更換對應的 Binder 綁定器即可，不需要修改任何應用邏輯，甚至可以任意改變中介軟體的類型而不需要修改一行程式。阿里巴巴提供了 RocketMQ 的 Binder 實現──spring-cloud-stream- binder-rocketmq。

8.2.2 Kotlin 整合 RocketMQ 實現訊息驅動

新增 Maven 子專案：chapter08-stream-rocketmq，這是一個以 RocketMQ 為基礎的 Spring Cloud Stream 應用，pom 檔案如下：

```xml
1.  <?xml version="1.0" encoding="UTF-8"?>
2.  <project xmlns="http://maven.apache.org/POM/4.0.0"
3.          xmlns:xsi="http://www.w3.org/2001/XMLSchema-instance"
4.          xsi:schemaLocation="http://maven.apache.org/POM/4.0.0 http://
    maven.apache.org/xsd/maven-4.0.0.xsd">
5.      <parent>
6.          <artifactId>kotlinspringboot</artifactId>
7.          <groupId>io.kang.kotlinspringboot</groupId>
8.          <version>0.0.1-SNAPSHOT</version>
9.      </parent>
10.     <modelVersion>4.0.0</modelVersion>
11.     <!-- 子專案名 -->
12.     <artifactId>chapter08-stream-rocketmq</artifactId>
```

```
13.    <dependencies>
14.        <!-- Spring Boot Web 相依套件 -->
15.        <dependency>
16.            <groupId>org.springframework.boot</groupId>
17.            <artifactId>spring-boot-starter-web</artifactId>
18.            <version>2.2.1.RELEASE</version>
19.        </dependency>
20.        <!-- Spring Cloud RocketMQ 相依套件 -->
21.        <dependency>
22.            <groupId>com.alibaba.cloud</groupId>
23.            <artifactId>spring-cloud-stream-binder-rocketmq</artifactId>
24.            <version>2.2.0.RELEASE</version>
25.        </dependency>
26.        <dependency>
27.            <groupId>com.fasterxml.jackson.module</groupId>
28.            <artifactId>jackson-module-kotlin</artifactId>
29.        </dependency>
30.        <dependency>
31.            <groupId>org.jetbrains.kotlin</groupId>
32.            <artifactId>kotlin-reflect</artifactId>
33.        </dependency>
34.        <dependency>
35.            <groupId>org.jetbrains.kotlin</groupId>
36.            <artifactId>kotlin-stdlib-jdk8</artifactId>
37.        </dependency>
38.        <dependency>
39.            <groupId>org.jetbrains.kotlinx</groupId>
40.            <artifactId>kotlinx-coroutines-core</artifactId>
41.            <version>1.3.2</version>
42.        </dependency>
43.    </dependencies>
44.    <build>
45.        <sourceDirectory>${project.basedir}/src/main/kotlin
    </sourceDirectory>
46.        <testSourceDirectory>${project.basedir}/src/test/kotlin
```

```
      </testSourceDirectory>
47.       <plugins>
48.           <plugin>
49.               <groupId>org.springframework.boot</groupId>
50.               <artifactId>spring-boot-maven-plugin</artifactId>
51.           </plugin>
52.           <plugin>
53.               <groupId>org.jetbrains.kotlin</groupId>
54.               <artifactId>kotlin-maven-plugin</artifactId>
55.               <configuration>
56.                   <args>
57.                       <arg>-Xjsr305=strict</arg>
58.                   </args>
59.                   <compilerPlugins>
60.                       <plugin>spring</plugin>
61.                       <plugin>jpa</plugin>
62.                   </compilerPlugins>
63.               </configuration>
64.               <dependencies>
65.                   <dependency>
66.                       <groupId>org.jetbrains.kotlin</groupId>
67.                       <artifactId>kotlin-maven-allopen</artifactId>
68.                       <version>${kotlin.version}</version>
69.                   </dependency>
70.                   <dependency>
71.                       <groupId>org.jetbrains.kotlin</groupId>
72.                       <artifactId>kotlin-maven-noarg</artifactId>
73.                       <version>${kotlin.version}</version>
74.                   </dependency>
75.               </dependencies>
76.           </plugin>
77.       </plugins>
78.   </build>
79. </project>
```

application.yml 檔案的內容如下：

```yaml
1.  server:
2.    port: 8082                          # 應用通訊埠
3.  spring:
4.    application:
5.      name: rocketmq-stream              # 應用名
6.    cloud:
7.      stream:
8.        rocketmq:
9.          binder:
10.            name-server: localhost:9876  #RocketMQ name-server 位址
11.          bindings:
12.            input1:
13.              consumer:
14.                orderly: true            # 順序消費
15.                tags: tagStr0            # 訊息 tag
16.            input2:
17.              consumer:
18.                orderly: false           # 不保證消費順序
19.                tags: tagStr1            # 訊息 tag
20.        bindings:
21.          output1:
22.            destination: test-topic      #topic 名稱
23.            content-type: text/plain     # 訊息格式
24.            producer:
25.              partitionKeyExpression: headers['partitionKey']  # 分區
26.              partitionCount: 2          # 分區個數
27.          input1:
28.            destination: test-topic      #topic 名稱
29.            content-type: text/plain     # 訊息格式
30.            group: test-consumer-group1  # 消費群組
31.            consumer:
32.              instance-index: 0          # 用哪個分區來接收訊息
33.              instance-count: 2          # 分區數
```

```
34.     input2:
35.        destination: test-topic        #topic 名稱
36.        content-type: text/plain       # 訊息格式
37.        group: test-consumer-group2    # 消費群組
38.        consumer:
39.           concurrency: 20             # 平行度
40.           instance-index: 1           #用哪個分區來接收訊息
41.           instance-count: 2           # 分區數
```

application.yml 定義了一個 output 和兩個 input。output 相當於生產者，input 相當於消費者。output1 向兩個分區發送訊息，分區根據 "partitionKey" 進行區分。input1 屬於消費群組 "test-consumer-group1"，接收分區 0 的訊息，接收 tag 是 "tagStr0" 的訊息。input2 屬於消費群組 "test-consumer-group2"，接收分區 1 的訊息，接收 tag 是 "tagStr1" 的訊息。這兩個消費群組都可以消費 "test-topic" 的訊息，相當於廣播模式。同一個消費群組有多個實例，共同使用（消費）topic 的訊息，每筆訊息只會被一個消費群組使用。

StreamApplication.kt 定義了一個啟動類別，植入自訂的輸入、輸出類別：

```
1.   // 開啟 Spring Cloud Stream Binding 註釋
2.   @SpringBootApplication
3.   @EnableBinding(MySource::class, MySink::class)
4.   class StreamApplication
5.   // 啟動函數
6.   fun main(args: Array<String>) {
7.       runApplication<StreamApplication>(*args)
8.   }
```

MySink.kt 定義了兩個輸入，即兩個消費者：

```
1.   interface MySink {
2.       // 輸入 1
3.       @Input("input1")
```

```
4.      fun input1(): SubscribableChannel
5.      // 輸入 2
6.      @Input("input2")
7.      fun input2(): SubscribableChannel
8.  }
```

MySource.kt 定義了一個輸出，即一個生產者：

```
1.  interface MySource {
2.      // 輸出 1
3.      @Output("output1")
4.      fun output1(): MessageChannel
5.  }
```

ReceiveService.kt 定義了兩個消費者的行為，input1 和 input2 收到訊息後
分別列印訊息內容：

```
1.  @Service
2.  class ReceiveService {
3.      // 監聽 input1，收到訊息，進行列印
4.      @StreamListener("input1")
5.      fun receiveInput1(receiveMsg: String) {
6.          println("input1 receive: $receiveMsg")
7.      }
8.      // 監聽 input2，收到訊息，進行列印
9.      @StreamListener("input2")
10.     fun receiveInput2(receiveMsg: String) {
11.         println("input2 receive: $receiveMsg")
12.     }
13. }
```

SteamController.kt 定義了三個介面，用於測試。"/send0" 介面向第 0 個分
區發送訊息，訊息的 tag 是 "tagStr0"。"/send1" 介面向第 1 個分區發送訊
息，訊息的 tag 是 "tagStr1"。"/send2" 介面向第 1 個分區發送訊息，訊息
的 tag 是 "tagStr2"。程式如下：

```
1.  @RestController
2.  class SteamController {
3.      @Autowired
4.      lateinit var mySource: MySource
5.      // 測試介面，向第 0 個分區發送 tag 是 tagStr0 的訊息
6.      @GetMapping("/send0/{id}")
7.      fun send(@PathVariable id: String) {
8.          val headers = HashMap<String, Any>()
9.          headers[MessageConst.PROPERTY_TAGS] = "tagStr0"
10.         headers["partitionKey"] = 0
11.         mySource.output1().send(MessageBuilder.createMessage("hello world:
    $id", MessageHeaders(headers)))
12.     }
13.     // 測試介面，向第 1 個分區發送 tag 是 tagStr1 的訊息
14.     @GetMapping("/send1/{id}")
15.     fun send1(@PathVariable id: String) {
16.         val headers = HashMap<String, Any>()
17.         headers[MessageConst.PROPERTY_TAGS] = "tagStr1"
18.         headers["partitionKey"] = 1
19.         mySource.output1().send(MessageBuilder.createMessage("hello world:
    $id", MessageHeaders(headers)))
20.     }
21.     // 測試介面，向第 1 個分區發送 tag 是 tagStr2 的訊息
22.     @GetMapping("/send2/{id}")
23.     fun send2(@PathVariable id: String) {
24.         val headers = HashMap<String, Any>()
25.         headers[MessageConst.PROPERTY_TAGS] = "tagStr2"
26.         headers["partitionKey"] = 1
27.         mySource.output1().send(MessageBuilder.createMessage("hello world:
    $id", MessageHeaders(headers)))
28.     }
29. }
```

呼叫 "/send0/1"，在主控台列印 "input1 receive: hello world: 1"，input1 成功
消費這筆訊息。呼叫 "/send1/1"，在主控台列印 "input2 receive: hello world:

1"，input2 成功消費這筆訊息。呼叫 "/send2/1"，在主控台沒有列印，因為 input2 只消費 tag 是 "tagStr1" 的訊息，這筆訊息的 tag 是 "tagStr2"，被過濾掉了。透過圖 8.5 可以看到，這筆訊息被 "test-consumer-group1" 和 "test-consumer-group2" 都過濾掉了。被 "test-consumer-group1" 過濾掉的原因是這筆訊息是第 1 個分區中的，而 "test-consumer-group1" 只消費發送到第 0 個分區的訊息。被 "test-consumer-group2" 過濾掉的原因是 "test-consumer-group2" 只消費 tag 是 "tagStr1" 的訊息。

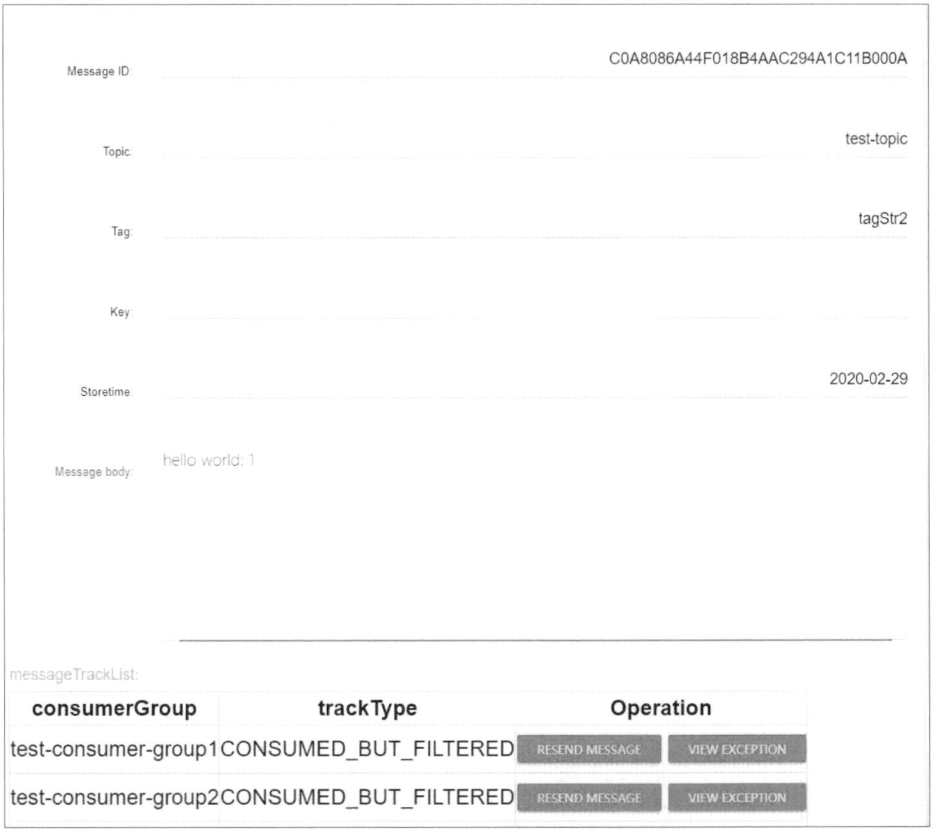

Message ID	C0A8086A44F018B4AAC294A1C11B000A
Topic	test-topic
Tag	tagStr2
Key	
Storetime	2020-02-29
Message body	hello world: 1

messageTrackList:

consumerGroup	trackType	Operation	
test-consumer-group1	CONSUMED_BUT_FILTERED	RESEND MESSAGE	VIEW EXCEPTION
test-consumer-group2	CONSUMED_BUT_FILTERED	RESEND MESSAGE	VIEW EXCEPTION

圖 8.5 tag=tagStr2 的訊息的消費結果

8.3 阿里物件雲端儲存

阿里物件雲端儲存是阿里巴巴提供的物件儲存 OSS（Object Storage Service）產品，方便使用者儲存圖片資源，並透過 URL 存取圖片。本節主要介紹 Kotlin 整合阿里雲 OSS 的 SDK 進行檔案上傳及下載開發的方法。

8.3.1 阿里物件雲端儲存介紹

為了更進一步地了解 OSS，這裡介紹以下幾個概念。

儲存空間（Bucket）是使用者儲存物件（Object）的容器，所有的物件都必須隸屬於某個儲存空間。儲存空間具有各種設定屬性，包含地域、存取權限、儲存類型等。使用者可以根據實際需求，建立不同類型的儲存空間來儲存不同的資料。同一個儲存空間的內部是扁平的，沒有檔案系統中的目錄等概念，所有的物件都直接隸屬於其對應的儲存空間。每個使用者可以擁有多個儲存空間。儲存空間的名稱在 OSS 範圍內必須是全域唯一的，一旦建立便無法修改名稱。儲存空間內部的物件數目沒有限制。

物件 / 檔案（Object）是 OSS 儲存資料的基本單元，也被稱為 OSS 的檔案。物件由詮譯資訊（Object Meta）、使用者資料（Data）和檔案名稱（Key）組成。物件由儲存空間內部唯一的 Key 來標識。物件詮譯資訊是一組鍵值對，表示物件的一些屬性，例如最後修改時間、物件大小等資訊，同時使用者也可以在詮譯資訊中儲存一些自訂的資訊。物件的生命週期是從上傳成功到被刪除為止。在整個生命週期內，只有透過追加上傳的物件可以繼續透過追加上傳寫入資料，以其他上傳方式上傳的物件內容無法編輯，可以透過重複上傳名稱相同的物件來覆蓋之前的物件。

地域（Region）表示 OSS 的資料中心所在的物理位置。使用者可以根據費用、請求來源等選擇合適的地域建立儲存空間。一般來說，距離使用者近的地域的存取速度更快。地域是在建立儲存空間的時候指定的，一旦指定儲存空間就不允許更改。該儲存空間中所有的物件都儲存在對應的資料中心，目前不支持對象等級的地域設定。

存取域名（Endpoint）表示 OSS 對外服務的存取域名。OSS 以 HTTP RESTful API 的形式對外提供服務，當存取不同地域的時候，需要不同的域名。透過內網和外網存取同一個地域所需要的存取域名也是不同的。

存取金鑰（AccessKey）是存取身份驗證中用到的 AccessKeyId 和 Access KeySecret。OSS 透過使用 AccessKeyId 和 AccessKeySecret 對稱加密的方法來驗證某個請求的發送者身份。AccessKeyId 用於標識使用者；AccessKeySecret 是使用者加密簽名字串和 OSS 用來驗證簽名字串的金鑰，必須保密。

物件操作在 OSS 上具有原子性，OSS 保障使用者一旦上傳完成，讀到的物件就是完整的，OSS 不會給使用者傳回一個「部分上傳成功」的物件。物件操作在 OSS 上同樣具有強一致性，使用者一旦收到一個上傳成功的回應，該上傳的物件就已經立即讀取，並且物件的容錯資料也同時寫成功。

OSS 採用資料容錯儲存機制，將每個物件的不同容錯儲存在同一個區域內多個設施的多個裝置上，以確保硬體故障時的資料可靠性和可用性。OSS 的容錯儲存機制支援兩個儲存設施平行處理損壞時，仍可維持資料不遺失。當資料存入 OSS 後，OSS 會檢測和修復遺失的容錯，確保資料可靠性和可用性。OSS 會週期性地透過驗證等方式驗證資料的完整性，及時發現因硬體故障等原因造成的資料損壞。當檢測到資料有部分損壞或遺失時，OSS 會利用容錯的資料進行重建並修復損壞的資料。

8.3.2 Kotlin 整合阿里物件雲端儲存

新增一個 Maven 子專案：chapter08-oss，透過 ossClient 存取阿里雲物件儲存。pom.xml 檔案如下：

```xml
1.  <?xml version="1.0" encoding="UTF-8"?>
2.  <project xmlns="http://maven.apache.org/POM/4.0.0"
3.          xmlns:xsi="http://www.w3.org/2001/XMLSchema-instance"
4.          xsi:schemaLocation="http://maven.apache.org/POM/4.0.0 http://
    maven.apache.org/xsd/maven-4.0.0.xsd">
5.      <parent>
6.          <artifactId>kotlinspringboot</artifactId>
7.          <groupId>io.kang.kotlinspringboot</groupId>
8.          <version>0.0.1-SNAPSHOT</version>
9.      </parent>
10.     <modelVersion>4.0.0</modelVersion>
11.     <!-- 子專案名 -->
12.     <artifactId>chapter08-oss</artifactId>
13.     <dependencies>
14.         <!-- Spring Boot Web 相依套件 -->
15.         <dependency>
16.             <groupId>org.springframework.boot</groupId>
17.             <artifactId>spring-boot-starter-web</artifactId>
18.             <version>2.2.1.RELEASE</version>
19.         </dependency>
20.         <!-- Spring Boot Cloud OSS 相依套件 -->
21.         <dependency>
22.             <groupId>com.alibaba.cloud</groupId>
23.             <artifactId>spring-cloud-starter-alicloud-oss</artifactId>
24.             <version>2.2.0.RELEASE</version>
25.         </dependency>
26.         <dependency>
27.             <groupId>com.fasterxml.jackson.module</groupId>
28.             <artifactId>jackson-module-kotlin</artifactId>
29.         </dependency>
```

```
30.        <dependency>
31.            <groupId>org.jetbrains.kotlin</groupId>
32.            <artifactId>kotlin-reflect</artifactId>
33.        </dependency>
34.        <dependency>
35.            <groupId>org.jetbrains.kotlin</groupId>
36.            <artifactId>kotlin-stdlib-jdk8</artifactId>
37.        </dependency>
38.        <dependency>
39.            <groupId>org.jetbrains.kotlinx</groupId>
40.            <artifactId>kotlinx-coroutines-core</artifactId>
41.            <version>1.3.2</version>
42.        </dependency>
43.        <dependency>
44.            <groupId>org.springframework.boot</groupId>
45.            <artifactId>spring-boot-starter-test</artifactId>
46.            <version>2.2.1.RELEASE</version>
47.            <scope>test</scope>
48.        </dependency>
49.    </dependencies>
50.    <build>
51.        <sourceDirectory>${project.basedir}/src/main/kotlin
    </sourceDirectory>
52.        <testSourceDirectory>${project.basedir}/src/test/kotlin
    </testSourceDirectory>
53.        <plugins>
54.            <plugin>
55.                <groupId>org.springframework.boot</groupId>
56.                <artifactId>spring-boot-maven-plugin</artifactId>
57.            </plugin>
58.            <plugin>
59.                <groupId>org.jetbrains.kotlin</groupId>
60.                <artifactId>kotlin-maven-plugin</artifactId>
61.                <configuration>
62.                    <args>
```

```
63.                    <arg>-Xjsr305=strict</arg>
64.                </args>
65.                <compilerPlugins>
66.                    <plugin>spring</plugin>
67.                    <plugin>jpa</plugin>
68.                </compilerPlugins>
69.            </configuration>
70.            <dependencies>
71.                <dependency>
72.                    <groupId>org.jetbrains.kotlin</groupId>
73.                    <artifactId>kotlin-maven-allopen</artifactId>
74.                    <version>${kotlin.version}</version>
75.                </dependency>
76.                <dependency>
77.                    <groupId>org.jetbrains.kotlin</groupId>
78.                    <artifactId>kotlin-maven-noarg</artifactId>
79.                    <version>${kotlin.version}</version>
80.                </dependency>
81.            </dependencies>
82.        </plugin>
83.    </plugins>
84.    </build>
85. </project>
```

application.yml 檔案如下所示，其中定義了 OSS 的造訪網址和金鑰資訊，將 access-key 和 secret-key 取代為自己的即可：

```
1.  server:
2.    port: 8082          # 應用通訊埠編號
3.  spring:
4.    application:
5.      name: oss-app      # 應用名
6.    cloud:
7.      alicloud:
8.        oss:
9.          endpoint: http://oss-cn-beijing.aliyuncs.com  #OSS 服務位址
```

```
10.      access-key:      # 阿里雲 access key
11.      secret-key:      # 阿里雲 secret key
```

OssApplication.kt 是一個啟動類別，啟動了一個 Spring Boot 應用：

```
1.  @SpringBootApplication
2.  class OssApplication
3.  // 啟動函數
4.  fun main(args: Array<String>) {
5.      runApplication<OssApplication>(*args)
6.  }
```

OssController.kt 定義了四個介面，用於測試向 OSS 上傳、下載檔案。
"/upload" 和 "/download" 使 用 ossClient 上 傳 及 下 載 oss-test.json 檔 案。
"upload2" 和 "/file-resource" 使用 Resource 方式上傳及下載 oss-test.json 檔
案。ossClient 通常用於操作大量檔案物件的場景，如果只需讀取少量檔
案，可以用 Resource 的形式獲得檔案物件。程式如下：

```
1.  @RestController
2.  class OssController {
3.      @Autowired
4.      lateinit var ossClient: OSS
5.      // 本機檔案位置
6.      @Value("classpath:/oss-test.json")
7.      lateinit var localFile: Resource
8.      // 遠端檔案位置
9.      @Value("oss://kcglobal/test/oss-test.json")
10.     lateinit var remoteFile: Resource
11.     // 測試介面，上傳檔案到 OSS
12.     @GetMapping("/upload")
13.     fun upload(): String {
14.         try {
15.             val stream = this.javaClass.classLoader.getResourceAsStream
    ("oss-test.json")
16.             ossClient.putObject("kcglobal", "test/oss-test.json", stream)
```

```
17.          } catch (e: Exception) {
18.              return "upload fail: " + e.message
19.          }
20.          return "upload success"
21.     }
22.     // 測試介面，從 OSS 讀取檔案
23.     @GetMapping("/file-resource")
24.     fun fileResource(): String {
25.          return try {
26.              "get file resource success. content: " +
27.                      StreamUtils.copyToString(remoteFile.inputStream,
     Charset.forName(CharEncoding.UTF_8))
28.          } catch (e: Exception) {
29.              "get resource fail: " + e.message
30.          }
31.     }
32.     // 測試介面，從 OSS 讀取檔案
33.     @GetMapping("/download")
34.     fun download(): String {
35.          return try {
36.              val ossObject = ossClient.getObject("kcglobal", "test/oss-
     test.json")
37.              "download success, content: " + IOUtils.readStreamAsString
     (ossObject.objectContent, CharEncoding.UTF_8)
38.          } catch (e: Exception) {
39.              "download fail: " + e.message
40.          }
41.     }
42.     // 測試介面，上傳檔案到 OSS
43.     @GetMapping("/upload2")
44.     fun uploadWithOutputStream(): String {
45.          try {
46.          (this.remoteFile as WritableResource)
47.                  .outputStream
48.                      .use { outputStream ->
```

```
49.                        localFile.inputStream.use { inputStream ->
    StreamUtils.copy(inputStream, outputStream)
50.                            }
51.                        }
52.            } catch (ex: Exception) {
53.                return "upload with outputStream failed"
54.            }
55.            return "upload success"
56.        }
57. }
```

OssImageController.kt 定義了兩個介面，用於測試上傳和下載圖片。
"image" 介面用於取得圖片的造訪網址，該位址有過期時間，"image/
upload" 介面用於上傳圖片：

```
1.  @RestController
2.  class OssImageController {
3.      @Autowired
4.      lateinit var ossClient: OSS
5.
6.      val imageExpireTime = 10 * 365 * 24 * 60 * 60 * 1000L;
7.      // 取得圖片的造訪網址
8.      @GetMapping("/image/{name}")
9.      fun getImageUrl(@PathVariable name: String): String? {
10.         val expiration = Date(Date().getTime() + imageExpireTime)
11.         val url = ossClient.generatePresignedUrl("kcglobal", "test/" +
    name, expiration)
12.         return url?.toString()
13.     }
14.     // 上傳圖片
15.     @GetMapping("/image/upload/{name}")
16.     fun uploadImage(@PathVariable name: String): String {
17.         var ret = ""
18.         var stream: InputStream ?= null
19.         try {
```

```
20.          val objectMetadata = ObjectMetadata()
21.          stream = this.javaClass.classLoader.getResourceAsStream
    ("images/$name")
22.          objectMetadata.contentLength = stream.available().toLong()
23.          objectMetadata.cacheControl = "no-cache";
24.          objectMetadata.setHeader("Pragma", "no-cache");
25.          objectMetadata.contentType = "image/jpg"
26.          objectMetadata.contentDisposition = "inline;filename=$name"
27.          val putObject = ossClient.putObject("kcglobal", "test/$name",
    stream, objectMetadata)
28.          ret = putObject.eTag
29.       } catch (e: IOException) {
30.          println("upload file to oss error name=$name")
31.       } finally {
32.          try{
33.              if(stream != null) {
34.                  stream.close()
35.              }
36.          } catch (e: Exception) {
37.              e.printStackTrace()
38.          }
39.       }
40.       return ret
41.    }
42. }
```

8.4 分散式任務排程

SchedulerX 是阿里巴巴集團中介軟體團隊開發的一款高性能、分散式的任務排程產品，在阿里巴巴內部具有廣泛的使用，經過集團內上千個業務應用、歷經多年打磨而成。截至 2016 年 6 月，在 SchedulerX 上每天平穩執行集團內的幾十萬個任務，完成每天幾億次的任務排程。本節介紹 Kotlin 整合 SchedulerX 的 SDK 進行定時任務開發的相關知識。

8.4.1 SchedulerX 介紹

SchedulerX 1.0 版本讓任務實現了分散式。SchedulerX 1.0 提供了自主運行維護管理的後台，讓使用者能透過頁面來設定、修改和管理定時任務，使用者只需在頁面上修改時間運算式，不需要重新發佈執行定時任務的業務應用。SchedulerX 1.0 還能管理任務執行的生命週期，從任務執行開始一直到任務執行結束都有記錄，使用者能看到每次任務執行的開始時間和結束時間，還能看到執行成功或失敗，SchedulerX 1.0 還會為使用者保留過去的執行記錄，讓使用者可以檢視定時任務歷史執行記錄。

此外，SchedulerX 1.0 能把一個執行耗時很長的定時任務拆分成多個子任務分片，分發到多台機器上並存執行，大幅減少了定時任務的執行時間。

SchedulerX 2.0 版本提供了完整的任務排程系統。SchedulerX 2.0（DTS）進一步提升了使用者體驗，除最佳化程式設計模型、減少使用者設定和程式介面之外，還新增了多項功能。SchedulerX 2.0（DTS）支援七種功能。

簡單 job 單機版是每次隨機選擇一台機器執行任務，即任務只執行在一台機器上。但是為了防止單點故障還得解決多機備份的問題，當一台機器當機的時候可以自動切換到其他正常執行的機器上去。

簡單 job 廣播版則是每次選擇所有機器同時執行任務，例如需要定時更新本機記憶體的場景，這時就需要每台機器同時更新記憶體。

平行計算 job 是將一個耗時很長的大任務拆分成多個小的子任務然後分發到多台機器去並存執行。

圖示計算（任務依賴），這個功能多用於在有業務資料依賴的多個任務之間按照嚴格先後循序執行的場景。例如兩個任務 A、B，其中 A 執行結束之後 B 才能開始執行。

指令稿 job 是指 shell、PHP、Python 等定時執行的指令稿任務，使用者
只需在 SchedulerX 2.0（DTS）管理後台設定要定時執行的 shell、php、
python 等指令即可，使用者不需要額外寫任何程式。

SchedulerX 2.0（DTS）的管理運行維護主控台提供任務設定管理，以及歷
史執行記錄查詢，還有完整的監控警告功能。任務沒有準時執行會給使用
者發送警告，任務執行超過預期的時間也會給使用者發送警告，任務執行
失敗了也會給使用者發送簡訊警告。

SchedulerX 2.0（DTS）還支援以 SchedulerX 2.0 為基礎的延伸開發，使用
者可以透過 SDK 中的 API 來建立、修改和刪除任務。

SchedulerX 2.0（DTS）還支援超大規模定時觸發器，使用者可以透過 API
建立千億量級的一次性定時觸發器，例如每條交易訂單建立的時候就在
SchedulerX 2.0 中建立一個定時觸發器，使用者透過設定這個觸發器的觸
發時間，使得每到觸發時間，事件通知交易系統就會提醒使用者確認收貨
逾時。

8.4.2 Kotlin 整合 SchedulerX

新增一個 Maven 子專案：chapter08-schedulerx，pom.xml 如下：

```xml
1.  <?xml version="1.0" encoding="UTF-8"?>
2.  <project xmlns="http://maven.apache.org/POM/4.0.0"
3.          xmlns:xsi="http://www.w3.org/2001/XMLSchema-instance"
4.          xsi:schemaLocation="http://maven.apache.org/POM/4.0.0 http://
    maven.apache.org/xsd/maven-4.0.0.xsd">
5.      <parent>
6.          <artifactId>kotlinspringboot</artifactId>
7.          <groupId>io.kang.kotlinspringboot</groupId>
8.          <version>0.0.1-SNAPSHOT</version>
9.      </parent>
```

```
10.     <modelVersion>4.0.0</modelVersion>
11.     <!-- 子專案名 -->
12.     <artifactId>chapter08-schedulerx</artifactId>
13.     <dependencies>
14.         <!-- Spring Cloud SchedulerX 相依套件 -->
15.         <dependency>
16.             <groupId>com.alibaba.cloud</groupId>
17.             <artifactId>spring-cloud-starter-alicloud-schedulerx
    </artifactId>
18.             <version>2.2.0.RELEASE</version>
19.         </dependency>
20.         <dependency>
21.             <groupId>com.fasterxml.jackson.module</groupId>
22.             <artifactId>jackson-module-kotlin</artifactId>
23.         </dependency>
24.         <dependency>
25.             <groupId>org.jetbrains.kotlin</groupId>
26.             <artifactId>kotlin-reflect</artifactId>
27.         </dependency>
28.         <dependency>
29.             <groupId>org.jetbrains.kotlin</groupId>
30.             <artifactId>kotlin-stdlib-jdk8</artifactId>
31.         </dependency>
32.         <dependency>
33.             <groupId>org.jetbrains.kotlinx</groupId>
34.             <artifactId>kotlinx-coroutines-core</artifactId>
35.             <version>1.3.2</version>
36.         </dependency>
37.     </dependencies>
38.     <build>
39.         <sourceDirectory>${project.basedir}/src/main/kotlin
    </sourceDirectory>
40.         <testSourceDirectory>${project.basedir}/src/test/kotlin
    </testSourceDirectory>
41.         <plugins>
```

```
42.        <plugin>
43.            <groupId>org.springframework.boot</groupId>
44.            <artifactId>spring-boot-maven-plugin</artifactId>
45.        </plugin>
46.        <plugin>
47.            <groupId>org.jetbrains.kotlin</groupId>
48.            <artifactId>kotlin-maven-plugin</artifactId>
49.            <configuration>
50.                <args>
51.                    <arg>-Xjsr305=strict</arg>
52.                </args>
53.                <compilerPlugins>
54.                    <plugin>spring</plugin>
55.                    <plugin>jpa</plugin>
56.                </compilerPlugins>
57.            </configuration>
58.            <dependencies>
59.                <dependency>
60.                    <groupId>org.jetbrains.kotlin</groupId>
61.                    <artifactId>kotlin-maven-allopen</artifactId>
62.                    <version>${kotlin.version}</version>
63.                </dependency>
64.                <dependency>
65.                    <groupId>org.jetbrains.kotlin</groupId>
66.                    <artifactId>kotlin-maven-noarg</artifactId>
67.                    <version>${kotlin.version}</version>
68.                </dependency>
69.            </dependencies>
70.        </plugin>
71.      </plugins>
72.    </build>
73. </project>
```

application.yml 檔案如下：

```
1. server:
```

```
2.    port: 8084              # 應用通訊埠編號
3.  spring:
4.    application:
5.      name: SCX-APP         # 應用名稱
6.    cloud:
7.      alicloud:
8.        scx:
9.          group-id: ***     #SchedulerX 的 group id
10.       edas:
11.         namespace: cn-test   #namespace 名稱
```

SpringBootApplication.kt 是一個 Spring Boot 啟動類別：

```
1.  @SpringBootApplication
2.  class ScxApplication
3.  // 啟動函數
4.  fun main(args: Array<String>) {
5.      runApplication<ScxApplication>(*args)
6.  }
```

SimpleTask.kt 定義了一個 job 處理類別，列印 "Hello World"：

```
1.  // job 處理類別
2.  class SimpleTask: ScxSimpleJobProcessor {
3.      override fun process(context: ScxSimpleJobContext?): ProcessResult {
4.          println("-----------Hello world---------------")
5.          return ProcessResult(true)
6.      }
7.  }
```

進入阿里雲的 SchedulerX 任務清單頁面，選擇上方的「測試」區域，點擊右上角的「新增 Job」，建立一個 job，如下所示：

```
1.  job 分組：測試—***-*-*-****
2.  job 處理介面：io.kang.schedule.SimpleTask
3.  類型：簡單 job 單機版
```

以上任務類型選擇了「簡單 job 單機版」，並且制定了 Cron 運算式為 "0 * * * * ?"，這表示，每過 1 分鐘，任務將被執行且只執行 1 次。

8.5 分散式交易

分散式交易是跨資料庫的交易問題。為了解決資料庫存取的瓶頸問題，切分資料庫是很常見的解決方案，不同使用者可能落在不同的資料庫裡，原來在一個函數庫裡的交易操作，現在變成了跨資料庫的交易操作。隨著業務的不斷增長，將業務不同模組中的服務拆分成微服務後，每個微服務都對應一個獨立的資料來源，資料來源可能位於不同的機房，同時呼叫多個微服務很難保障同時成功，因此會產生跨服務分散式交易問題。本節介紹使用 Kotlin 整合 Seata 分散式架構進行分散式交易開發的相關知識。

8.5.1 分散式交易介紹

常見的分散式解決方案有：兩階段提交（2pc）、TCC 和交易訊息。

兩階段提交協定有關兩種角色：一個是交易協調者（coordinator），負責協調多個參與者進行交易投票及提交（回覆）；另一個是多個交易參與者（participants），即本機交易執行者。處理步驟有以下兩個。

投票階段（voting phase）：協調者將通知交易參與者準備提交或取消交易，然後進入表決過程。參與者將告知協調者自己的決策——同意（交易參與者本機交易執行成功，但未提交）或取消（本機交易執行故障）。

提交階段（commit phase）：收到參與者的通知後，協調者再向參與者發出通知，根據回饋情況決定各參與者提交還是回覆。

兩階段提交協定的實現很簡單，但是需要資料庫，需要 XA（eXtended Architecture）的強一致性支援，鎖粒度大，效能差。

TCC 將交易提交分為 Try - Confirm - Cancel 三個操作。其和兩階段提交有點類似，Try 為第一階段，Confirm - Cancel 為第二階段，是一種應用層面侵入業務的兩階段提交。它不依賴 RM（資源管理員）對分散式交易的支援，而是透過對業務邏輯的分解來實現分散式交易。

交易訊息更偏好達成分散式交易的最後一致性，分散式交易的提交或回覆取決於交易發起方的業務需求，如 A 給 B 匯了款並且成功了，那麼下游業務 B 的錢款一定要增加這種場景，或某使用者完成訂單，使用者積分一定要增加這種場景。

Seata 是阿里巴巴開放原始碼的分散式交易解決方案，對業務侵入度小，即減少技術架構上的微服務化所帶來的分散式交易問題對業務的侵入；高性能，即減少分散式交易解決方案所帶來的效能消耗。Seata 中有兩種分散式交易實現方案，AT 及 TCC。AT 模式主要關注多資料庫存取的資料一致性，當然也包含多服務下的多資料庫資料存取的一致性問題。TCC 模式主要關注業務拆分，在水平擴充業務資源時，解決微服務間呼叫的一致性問題。

Seata AT 模式是基於 XA 交易演進而來的分散式交易中介軟體，XA 是一個以資料庫實現為基礎的分散式交易協定，本質上和兩階段提交一樣，需要資料庫支援，MySQL 5.6 以上版本支援 XA 協定，其他資料庫如 Oracle、DB2 也實現了 XA 介面。Seata AT 模式中的各角色如圖 8.6 所示。

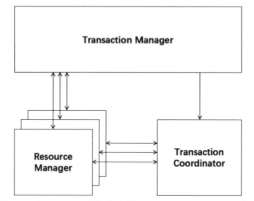

圖 8.6 Seata AT 模式中的 TC、RM、TM 示意圖

Transaction Coordinator（TC），交易協調器，維護全域交易的執行狀態，負責協調並驅動全域交易的提交或回覆。Transaction Manager（TM），控制全域交易的邊界，負責開啟一個全域交易，並最後發起全域提交或全域回覆的決議。Resource Manager（RM），控制分支交易，負責分支註冊、狀態匯報，並接收交易協調器的指令及驅動分支（本機）交易的提交和回覆。

第一階段，Seata 的 JDBC 資料來源代理透過對業務 SQL 的解析，把業務資料更新前後的資料映像檔組織成回覆記錄檔，利用本機交易的 ACID 特性，將業務資料的更新和回覆記錄檔寫入同一個本機交易中並提交。這樣，可以確保提交的任何業務資料的更新一定有對應的回覆記錄檔存在。以這樣的機制為基礎，分支的本機交易便可以在全域交易的第一階段提交，並馬上釋放本機交易鎖定的資源。Seata 和 XA 交易的不同之處是，兩階段提交通常對資源的鎖定需要持續到第二階段實際的提交或回覆操作，而有了回覆記錄檔之後，可以在第一階段釋放對資源的鎖定，降低了鎖定範圍，加強了效率，即使第二階段發生異常需要回覆，只需找到 undolog 中對應的資料並反解析成 SQL 即可達到回覆的目的。同時，Seata 透過代理資料來源將業務 SQL 的執行解析成 undolog 來與業務資料的更新同時寫入資料庫，達到了對業務無侵入的效果。

第二階段，如果決議是全域提交，此時分支交易已經完成提交，不需要同步協調處理（只需非同步清理回覆記錄檔），第二階段可以非常快速地完成。如果決議是全域回覆，RM 收到協調器發來的回覆請求，透過 XID 和 Branch ID 找到對應的回覆記錄檔記錄，透過回覆記錄產生反向的更新 SQL 並執行，以完成分支的回覆操作。

Seata 還針對 TCC 做了轉換相容，支援 TCC 交易方案，使用侵入業務上的補償及交易管理員的協調來達到全域交易的一起提交及回覆。

8.5.2 Kotlin 整合 Seata

新增一個 Maven 子專案：chapter08-seata，這是一個服務提供者。pom 檔案如下：

```xml
1.  <?xml version="1.0" encoding="UTF-8"?>
2.  <project xmlns="http://maven.apache.org/POM/4.0.0"
3.          xmlns:xsi="http://www.w3.org/2001/XMLSchema-instance"
4.          xsi:schemaLocation="http://maven.apache.org/POM/4.0.0 http://
    maven.apache.org/xsd/maven-4.0.0.xsd">
5.      <parent>
6.          <artifactId>kotlinspringboot</artifactId>
7.          <groupId>io.kang.kotlinspringboot</groupId>
8.          <version>0.0.1-SNAPSHOT</version>
9.      </parent>
10.     <modelVersion>4.0.0</modelVersion>
11.     <!-- 子專案名 -->
12.     <artifactId>chapter08-seata</artifactId>
13.     <dependencies>
14.         <dependency>
15.             <groupId>com.alibaba.cloud</groupId>
16.             <artifactId>spring-cloud-starter-alibaba-nacos-discovery
    </artifactId>
17.             <version>2.1.1.RELEASE</version>
```

```
18.    </dependency>
19.    <!-- Spring Cloud Seata 相依套件 -->
20.    <dependency>
21.        <groupId>com.alibaba.cloud</groupId>
22.        <artifactId>spring-cloud-starter-alibaba-seata</artifactId>
23.        <version>2.1.1.RELEASE</version>
24.    </dependency>
25.    <!-- Spring Boot Web 相依套件 -->
26.    <dependency>
27.        <groupId>org.springframework.boot</groupId>
28.        <artifactId>spring-boot-starter-web</artifactId>
29.        <version>2.1.1.RELEASE</version>
30.    </dependency>
31.    <!-- Lombok 相依套件 -->
32.    <dependency>
33.        <groupId>org.projectlombok</groupId>
34.        <artifactId>lombok</artifactId>
35.    </dependency>
36.    <!-- MyBatis 相依套件 -->
37.    <dependency>
38.        <groupId>org.mybatis.spring.boot</groupId>
39.        <artifactId>mybatis-spring-boot-starter</artifactId>
40.        <version>2.1.1</version>
41.    </dependency>
42.    <!-- MySQL 驅動 -->
43.    <dependency>
44.        <groupId>mysql</groupId>
45.        <artifactId>mysql-connector-java</artifactId>
46.    </dependency>
47.    <dependency>
48.        <groupId>com.fasterxml.jackson.module</groupId>
49.        <artifactId>jackson-module-kotlin</artifactId>
50.    </dependency>
51.    <dependency>
52.        <groupId>org.jetbrains.kotlin</groupId>
```

```
53.            <artifactId>kotlin-reflect</artifactId>
54.        </dependency>
55.        <dependency>
56.            <groupId>org.jetbrains.kotlin</groupId>
57.            <artifactId>kotlin-stdlib-jdk8</artifactId>
58.        </dependency>
59.        <dependency>
60.            <groupId>org.jetbrains.kotlinx</groupId>
61.            <artifactId>kotlinx-coroutines-core</artifactId>
62.            <version>1.3.2</version>
63.        </dependency>
64.        <dependency>
65.            <groupId>org.springframework.boot</groupId>
66.            <artifactId>spring-boot-starter-test</artifactId>
67.            <scope>test</scope>
68.        </dependency>
69.    </dependencies>
70.    <build>
71.        <sourceDirectory>${project.basedir}/src/main/kotlin
    </sourceDirectory>
72.        <testSourceDirectory>${project.basedir}/src/test/kotlin
    </testSourceDirectory>
73.        <plugins>
74.            <plugin>
75.                <groupId>org.springframework.boot</groupId>
76.                <artifactId>spring-boot-maven-plugin</artifactId>
77.            </plugin>
78.            <plugin>
79.                <groupId>org.jetbrains.kotlin</groupId>
80.                <artifactId>kotlin-maven-plugin</artifactId>
81.                <configuration>
82.                    <args>
83.                        <arg>-Xjsr305=strict</arg>
84.                    </args>
85.                    <compilerPlugins>
```

```
86.                        <plugin>spring</plugin>
87.                        <plugin>jpa</plugin>
88.                    </compilerPlugins>
89.                </configuration>
90.                <dependencies>
91.                    <dependency>
92.                        <groupId>org.jetbrains.kotlin</groupId>
93.                        <artifactId>kotlin-maven-allopen</artifactId>
94.                        <version>${kotlin.version}</version>
95.                    </dependency>
96.                    <dependency>
97.                        <groupId>org.jetbrains.kotlin</groupId>
98.                        <artifactId>kotlin-maven-noarg</artifactId>
99.                        <version>${kotlin.version}</version>
100.                   </dependency>
101.               </dependencies>
102.           </plugin>
103.       </plugins>
104.   </build>
105. </project>
```

application.yml 定義如下,定義了資料庫連接資訊,Nacos 服務註冊中心位址,Seata 交易服務群組:

```
1.  server:
2.    port: 8085                              # 應用通訊埠編號
3.  spring:
4.    application:
5.      name: sca-provider                    # 應用名
6.    datasource:
7.      driver-class-name: com.mysql.jdbc.Driver # 資料庫驅動
8.      url: jdbc:mysql://127.0.0.1:3306/video?characterEncoding=
    utf-8&serverTimezone=UTC                 # 資料庫連接位址
9.      username: root                        # 資料庫連接使用者名稱
10.     password: 123456                      # 資料庫連接密碼
```

```
11.   cloud:
12.     nacos:
13.       discovery:
14.         server-addr: 127.0.0.1:8848          #Nacos 註冊中心位址
15.       alibaba:
16.         seata:
17.           tx-service-group: sca-provider-group # 交易服務群組名稱
18. mybatis:
19.   mapper-locations: classpath:mapper/*.xml      # 資料庫操作的相關 XML 檔案
      所在位置
```

registry.conf 定義了 Seata 設定資訊儲存方式，可以儲存在本機檔案、
Nacos、Eureka、Redis、Zookeeper、Etcd、Sofa 中，這裡用本機檔案方式
儲存設定資訊：

```
1.   registry {
2.     # file、Nacos、Eureka、Redis、Zookeeper、Consul、Etcd3、Sofa
3.     # 採用檔案方式，檔案名稱是 file.conf
4.     type = "file"
5.     file {
6.       name = "file.conf"
7.     }
8.   }
9.   config {
10.    # file、Nacos、Apollo、Zookeeper、Consul、Etcd3、Spring Cloud Config
11.    # 採用檔案方式，檔案名稱是 file.conf
12.    type = "file"
13.    file {
14.      name = "file.conf"
15.    }
16. }
```

file.conf 定義了 Seata 的設定資訊，vgroup_mapping 的 sca-provider-group
和 application.yml 定義的 tx-service-group 保持一致，disableGlobalTransaction
用於控制是否開啟全域交易：

```
1.  transport {
2.    # 採用 TCP 通訊協定
3.    type = "TCP"
4.    # 服務端使用 NIO
5.    server = "NIO"
6.    # 開啟心跳
7.    heartbeat = true
8.    # 允許用戶端批次發送請求
9.    enableClientBatchSendRequest = true
10.   # Netty 執行緒工廠設定
11.   threadFactory {
12.     bossThreadPrefix = "NettyBoss"
13.     workerThreadPrefix = "NettyServerNIOWorker"
14.     serverExecutorThread-prefix = "NettyServerBizHandler"
15.     shareBossWorker = false
16.     clientSelectorThreadPrefix = "NettyClientSelector"
17.     clientSelectorThreadSize = 1
18.     clientWorkerThreadPrefix = "NettyClientWorkerThread"
19.     # Netty boss 執行緒大小
20.     bossThreadSize = 1
21.     # 工作執行緒大小
22.     workerThreadSize = "default"
23.   }
24.   shutdown {
25.     # 服務銷毀後，等待 3 秒
26.     wait = 3
27.   }
28.   # 序列化方式
29.   serialization = "seata"
30.   # 不壓縮
31.   compressor = "none"
32. }
33. service {
34.   # 交易服務群組對映
35.   vgroup_mapping.sca-provider-group = "default"
```

```
36.    # 當 registry.type=file 時，設定該屬性，不要設定多個位址
37.    default.grouplist = "127.0.0.1:8091"
38.    # 降級，目前不支援
39.    enableDegrade = false
40.    # 開啟 Seata 交易
41.    disableGlobalTransaction = false
42. }
43. client {
44.    # 資源管理相關設定
45.    rm {
46.      asyncCommitBufferLimit = 10000
47.      lock {
48.        retryInterval = 10
49.        retryTimes = 30
50.        retryPolicyBranchRollbackOnConflict = true
51.      }
52.      reportRetryCount = 5
53.      tableMetaCheckEnable = false
54.      reportSuccessEnable = false
55.    }
56.    # 交易管理相關設定
57.    tm {
58.      commitRetryCount = 5
59.      rollbackRetryCount = 5
60.    }
61.    undo {
62.      dataValidation = true
63.      logSerialization = "jackson"
64.      logTable = "undo_log"
65.    }
66.    log {
67.      exceptionRate = 100
68.    }
69. }
70. support {
```

```
71.    # 不允許自動代理資料來源
72.    spring.datasource.autoproxy = false
73. }
```

ProviderApp.kt 是啟動類別，啟動了一個 Spring Boot 服務，排除了 Spring Boot 附帶的資料來源設定類別，採用自訂的資料來源設定類別：

```
1.  // 排除資料來自動設定類別，開啟 Eureka 用戶端註釋
2.  @SpringBootApplication(exclude = [DataSourceAutoConfiguration::class])
3.  @EnableDiscoveryClient
4.  @MapperScan("io.kang.provider.mapper")
5.  class ProviderApp
6.  // 啟動函數
7.  fun main(args: Array<String>) {
8.      runApplication<ProviderApp>(*args)
9.  }
```

DataSourceConfiguration.kt 定義了 Seata 交易使用的資料來源，資料庫連接池用 Druid 載入 MyBatis 的 mapper 和設定：

```
1.  @Configuration
2.  @EnableConfigurationProperties(MybatisProperties::class)
3.  class DataSourceConfiguration {
4.      // 初始化資料來源
5.      @Bean
6.      @ConfigurationProperties(prefix = "spring.datasource")
7.      fun dataSource(): DataSource {
8.          return DruidDataSource()
9.      }
10.     // 初始化資料來源代理
11.     @Bean
12.     @Primary
13.     fun dataSourceProxy(dataSource: DataSource): DataSourceProxy {
14.         return DataSourceProxy(dataSource)
15.     }
16.     // 初始化 sqlSessionFactory
```

```
17.     @Bean
18.     fun sqlSessionFactoryBean(dataSourceProxy: DataSourceProxy,
19.                   mybatisProperties: MybatisProperties):
    SqlSessionFactoryBean {
20.         val bean = SqlSessionFactoryBean()
21.         bean.setDataSource(dataSourceProxy)
22.         val resolver = PathMatchingResourcePatternResolver()
23.         try {
24.             val locations = resolver.getResources(mybatisProperties.
    mapperLocations[0])
25.             bean.setMapperLocations(*locations)
26.             if (StringUtils.isNotBlank(mybatisProperties.configLocation)) {
27.                 val resources = resolver.getResources(mybatisProperties.
    configLocation)
28.                 bean.setConfigLocation(resources[0])
29.             }
30.         } catch (e: IOException) {
31.             e.printStackTrace()
32.         }
33.         return bean
34.     }
35. }
```

TbUserMapper.kt 定義了一個 mapper 介面，用於向 tb_user 表插入記錄：

```
1.  // mapper 介面
2.  interface TbUserMapper {
3.      fun insert(record: TbUser): Int
4.  }
```

TbUser.kt 定義了一個 TbUser 實體：

```
1.  // 實體類別
2.  @Data
3.  class TbUser : Serializable {
4.      var id: Int? = null
```

```
5.     var name: String? = null
6.     var age: Int? = null
7.  }
```

ProviderController.kt 提供了一個介面 "/add/user"，向資料庫插入一筆
TbUser 記錄：

```
1.  @RestController
2.  class ProviderController {
3.      @Autowired
4.      lateinit var userMapper: TbUserMapper
5.      // 測試介面
6.      @PostMapping("/add/user")
7.      fun add(@RequestBody user: TbUser) {
8.          println("add user: $user")
9.          user.name = "provider"
10.         userMapper.insert(user)
11.     }
12. }
```

新增一個 Maven 子專案：chapter08-seata-consumer，這是一個消費方，呼
叫 chapter08-seata 定義的 "/add/user" 介面。pom 檔案如下：

```
1.  <?xml version="1.0" encoding="UTF-8"?>
2.  <project xmlns="http://maven.apache.org/POM/4.0.0"
3.          xmlns:xsi="http://www.w3.org/2001/XMLSchema-instance"
4.          xsi:schemaLocation="http://maven.apache.org/POM/4.0.0 http://
    maven.apache.org/xsd/maven-4.0.0.xsd">
5.      <parent>
6.          <artifactId>kotlinspringboot</artifactId>
7.          <groupId>io.kang.kotlinspringboot</groupId>
8.          <version>0.0.1-SNAPSHOT</version>
9.      </parent>
10.     <modelVersion>4.0.0</modelVersion>
11.     <!-- 子專案名 -->
12.     <artifactId>chapter08-seata-consumer</artifactId>
```

```
13.    <dependencies>
14.        <dependency>
15.            <groupId>com.alibaba.cloud</groupId>
16.            <artifactId>spring-cloud-starter-alibaba-nacos-discovery
    </artifactId>
17.            <version>2.1.1.RELEASE</version>
18.        </dependency>
19.        <!-- Spring Cloud Seata 相依套件 -->
20.        <dependency>
21.            <groupId>com.alibaba.cloud</groupId>
22.            <artifactId>spring-cloud-starter-alibaba-seata</artifactId>
23.            <version>2.1.1.RELEASE</version>
24.        </dependency>
25.        <!-- Spring Boot Actuator 相依套件 -->
26.        <dependency>
27.            <groupId>org.springframework.boot</groupId>
28.            <artifactId>spring-boot-actuator</artifactId>
29.            <version>2.2.2.RELEASE</version>
30.        </dependency>
31.        <!-- Spring Boot Web 相依套件 -->
32.        <dependency>
33.            <groupId>org.springframework.boot</groupId>
34.            <artifactId>spring-boot-starter-web</artifactId>
35.        </dependency>
36.        <!-- Spring Cloud Feign 相依套件 -->
37.        <dependency>
38.            <groupId>org.springframework.cloud</groupId>
39.            <artifactId>spring-cloud-starter-openfeign</artifactId>
40.            <version>2.2.2.RELEASE</version>
41.        </dependency>
42.        <!-- Lombok 相依套件 -->
43.        <dependency>
44.            <groupId>org.projectlombok</groupId>
45.            <artifactId>lombok</artifactId>
46.        </dependency>
```

```
47.          <!-- MyBatis 相依套件 -->
48.      <dependency>
49.          <groupId>org.mybatis.spring.boot</groupId>
50.          <artifactId>mybatis-spring-boot-starter</artifactId>
51.          <version>2.1.1</version>
52.      </dependency>
53.          <!-- 資料庫連接相依套件 -->
54.      <dependency>
55.          <groupId>mysql</groupId>
56.          <artifactId>mysql-connector-java</artifactId>
57.      </dependency>
58.      <dependency>
59.          <groupId>com.fasterxml.jackson.module</groupId>
60.          <artifactId>jackson-module-kotlin</artifactId>
61.      </dependency>
62.      <dependency>
63.          <groupId>org.jetbrains.kotlin</groupId>
64.          <artifactId>kotlin-reflect</artifactId>
65.      </dependency>
66.      <dependency>
67.          <groupId>org.jetbrains.kotlin</groupId>
68.          <artifactId>kotlin-stdlib-jdk8</artifactId>
69.      </dependency>
70.      <dependency>
71.          <groupId>org.jetbrains.kotlinx</groupId>
72.          <artifactId>kotlinx-coroutines-core</artifactId>
73.          <version>1.3.2</version>
74.      </dependency>
75.    </dependencies>
76.    <build>
77.      <sourceDirectory>${project.basedir}/src/main/kotlin
   </sourceDirectory>
78.      <testSourceDirectory>${project.basedir}/src/test/kotlin
   </testSourceDirectory>
79.      <plugins>
```

```
80.              <plugin>
81.                  <groupId>org.springframework.boot</groupId>
82.                  <artifactId>spring-boot-maven-plugin</artifactId>
83.              </plugin>
84.              <plugin>
85.                  <groupId>org.jetbrains.kotlin</groupId>
86.                  <artifactId>kotlin-maven-plugin</artifactId>
87.                  <configuration>
88.                      <args>
89.                          <arg>-Xjsr305=strict</arg>
90.                      </args>
91.                      <compilerPlugins>
92.                          <plugin>spring</plugin>
93.                          <plugin>jpa</plugin>
94.                      </compilerPlugins>
95.                  </configuration>
96.                  <dependencies>
97.                      <dependency>
98.                          <groupId>org.jetbrains.kotlin</groupId>
99.                          <artifactId>kotlin-maven-allopen</artifactId>
100.                         <version>${kotlin.version}</version>
101.                     </dependency>
102.                     <dependency>
103.                         <groupId>org.jetbrains.kotlin</groupId>
104.                         <artifactId>kotlin-maven-noarg</artifactId>
105.                         <version>${kotlin.version}</version>
106.                     </dependency>
107.                 </dependencies>
108.             </plugin>
109.         </plugins>
110.     </build>
111. </project>
```

application.yml 定義了資料庫設定、Nacos 服務註冊中心位址、Seata 交易
服務群組：

```
1.  server:
2.    port: 8086                              # 應用通訊埠編號
3.  spring:
4.    datasource:
5.      driver-class-name: com.mysql.jdbc.Driver  # 資料庫驅動類別
6.      url: jdbc:mysql://127.0.0.1:3306/video?characterEncoding=
    utf-8&serverTimezone=UTC                  # 資料庫連接位址
7.      username: root                        # 資料庫連接使用者名稱
8.      password: 123456                      # 資料庫連接密碼
9.    cloud:
10.     nacos:
11.       discovery:
12.         server-addr: 127.0.0.1:8848       #Nacos 服務註冊中心位址
13.       alibaba:
14.         seata:
15.         # Seata 交易群組名稱，對應 file.conf 檔案中的 vgroup_mapping.sca-
    customer-seata- tx-service-group
16.           tx-service-group: sca-customer-group
17.    application:
18.      name: sca-customer        # 應用名稱
19. mybatis:
20.    mapper-locations: classpath:mapper/*.xml   # 資料庫操作相關的 XML 檔案
    所在的位置
```

file.conf、registry.conf 和 chapter08-seata 的定義基本一致，區別在於，file.conf 的以下設定項目，vgroup_mapping 的 sca-customer-group 和 application.yml 中定義的 seata-tx-service-group 保持一致：

```
1.  service {
2.    # 交易服務群組對映
3.    vgroup_mapping.sca-customer-group = "default"
4.  }
```

CustomerApp.kt 定義了一個啟動類別，排除 Spring Boot 附帶的資料來源設定類別：

```kotlin
1.  // 掃描 @FeignClient 註釋
2.  @SpringBootApplication(exclude = [DataSourceAutoConfiguration::class])
3.  @EnableDiscoveryClient
4.  @EnableFeignClients
5.  @MapperScan("io.kang.consumer.mapper")
6.  class CustomerApp
7.  // 啟動函數
8.  fun main(args: Array<String>) {
9.      runApplication<CustomerApp>(*args)
10. }
```

ProviderFeignService.kt 透過 Feign 方式存取 "chapter08-seata" 定義的 "/add/user" 介面：

```kotlin
1.  //Feign 介面
2.  @FeignClient(value = "sca-provider")
3.  interface ProviderFeignService {
4.      @PostMapping("/add/user")
5.      fun add(@RequestBody user: TbUser)
6.  }
```

UserController.kt 定義了一個介面 "/seata/user/add"，透過呼叫 "/user/add" 介面，用本機方法 localSave 向 tb_user 表插入一筆 TbUser 記錄。使用 @GlobalTransactional 開啟 Seata 分散式交易，當呼叫 "/user/add" 介面後拋出例外，chapter08-seata 的 "/user/add" 介面回覆，本機方法 localSave 也回覆，資料庫中沒有任何記錄。chapter08-seata 的 "/user/add" 介面、本機方法 localSave 分別向 tb_user 插入記錄，由於是分散式交易，出現異常可回覆，避免了資料不一致。

將設定檔 file.conf 中的 disableGlobalTransaction 設定為 true，關閉分散式交易，再次呼叫 "seata/user/add"，tb_user 表中出現兩筆記錄，表示交易沒有回覆。

```
1.  @RestController
2.  class UserController {
3.      @Autowired
4.      lateinit var userMapper: TbUserMapper
5.
6.      @Autowired
7.      lateinit var providerFeignService: ProviderFeignService
8.      // 測試介面，測試分散式交易
9.      @PostMapping("/seata/user/add")
10.     @GlobalTransactional(rollbackFor = [Exception::class])// 開啟全域交易
11.     fun add(@RequestBody user: TbUser) {
12.         println("globalTransactional begin, Xid: ${RootContext.getXID()}")
13.         // 使用本機方法儲存 user
14.         localSave(user)
15.         // 遠端呼叫介面儲存 user
16.         providerFeignService.add(user)
17.         // 拋出例外，測試 Seata 交易
18.         throw RuntimeException()
19.     }
20.     private fun localSave(user: TbUser) {
21.         user.name = "customer"
22.         userMapper.insert(user)
23.     }
24. }
```

圖 8.7 和圖 8.8 所示的兩個模組都出現了回覆記錄檔："Branch Rollbacked result: PhaseTwo_ Rollbacked"。

```
load [io.seata.rm.datasource.undo.parser.ProtostuffUndoLogParser] class fail. io/protostuff/runtime/Delegate
load UndoLogParser[jackson] extension by class[io.seata.rm.datasource.undo.parser.JacksonUndoLogParser]
Flipping property: sca-provider.ribbon.ActiveConnectionsLimit to use NEXT property: niws.loadbalancer.availabilityFilteringRule.activeConnectionsLimit = 21474
Client: sca-provider instantiated a LoadBalancer: DynamicServerListLoadBalancer:{NFLoadBalancer:name=sca-provider,current list of Servers=[],Load balancer sta
Using serverListUpdater PollingServerListUpdater
Flipping property: sca-provider.ribbon.ActiveConnectionsLimit to use NEXT property: niws.loadbalancer.availabilityFilteringRule.activeConnectionsLimit = 21474
DynamicServerListLoadBalancer for client sca-provider initialized: DynamicServerListLoadBalancer:{NFLoadBalancer:name=sca-provider,current list of Servers=[19
tion failure:0;    Total blackout seconds:0;    First connection made:Thu Jan 01 08:00:00 CST 1970;    First connection made: Thu Jan 01 08:00:00 CST 1970;    Act
onMessage:xid=192.168.126.1:8091:2045610353,branchId=2045610356,branchType=AT,resourceId=jdbc:mysql://127.0.0.1:3306/video,applicationData=null
Branch Rollbacking: 192.168.126.1:8091:2045610353 2045610356 jdbc:mysql://127.0.0.1:3306/video
Could not found property transaction.undo.data.validation, try to use default value instead.
Flipping property: sca-provider.ribbon.ActiveConnectionsLimit to use NEXT property: niws.loadbalancer.availabilityFilteringRule.activeConnectionsLimit = 21474
xid 192.168.126.1:8091:2045610353 branch 2045610356, undo_log deleted with GlobalFinished
Branch Rollbacked result: PhaseTwo_Rollbacked
[192.168.126.1:8091:2045610353] rollback status:Rollbacked
Servlet.service() for servlet [dispatcherServlet] in context with path [] threw exception [Request processing failed; nested exception is java.lang.RuntimeExc
```

圖 8.7 chapter08-seata-consumer 回覆記錄檔

```
Could not found property client.report.retry.count, try to use default value instead.
Could not found property client.lock.retry.policy.branch-rollback-on-conflict, try to use default value instead.
Could not found property client.lock.retry.internal, try to use default value instead.
Could not found property client.lock.retry.times, try to use default value instead.
load LoadBalance[null] extension by class[io.seata.discovery.loadbalance.RandomLoadBalance]
Could not found property transaction.undo.log.table, try to use default value instead.
Could not found property transaction.undo.log.serialization, try to use default value instead.
load [io.seata.rm.datasource.undo.parser.ProtostuffUndoLogParser] class fail. io/protostuff/runtime/RuntimeEnv
load UndoLogParser[jackson] extension by class[io.seata.rm.datasource.undo.parser.JacksonUndoLogParser]
onMessage:xid=192.168.126.1:8091:2045610353,branchId=2045610362,branchType=AT,resourceId=jdbc:mysql://127.0.0.1:3306/video,applicationData=null
Branch Rollbacking: 192.168.126.1:8091:2045610353 2045610362 jdbc:mysql://127.0.0.1:3306/video
Could not found property transaction.undo.data.validation, try to use default value instead.
xid 192.168.126.1:8091:2045610353 branch 2045610362, undo_log deleted with GlobalFinished
Branch Rollbacked result: PhaseTwo_Rollbacked
```

圖 8.8　chapter08-seata 回覆記錄檔

8.6 **Spring Cloud Dubbo**

Dubbo 是阿里巴巴內部的 SOA 服務化治理方案的核心架構，每天為 2000
多個服務提供多於 30 億次的造訪支援，並被廣泛應用於阿里巴巴集團的
各成員網站。Dubbo 自 2011 年開放原始碼後，已被許多非阿里系公司使
用，目前是 Apache 的頂級專案。本節介紹使用 Kotlin 整合 Spring Cloud
Dubbo 進行微服務開發的方法。

8.6.1 **Dubbo 介紹**

Dubbo 在中文社群擁有極大的使用者群，大家希望在使用 Dubbo 的同時享
受 Spring Cloud 的生態系統，因而出現了各式各樣的整合方案。但是因為
服務中心不同，各種整合方案並不是那麼自然，直到 Spring Cloud Alibaba
這個專案出現，由官方提供了 Nacos 服務註冊中心，才將這個問題完美解
決，並且提供了 Dubbo 和 Spring Cloud 整合的方案 Dubbo Spring Cloud。

Dubbo Spring Cloud 建置在原生的 Spring Cloud 之上，其在服務治理方面
的能力可被認為是 Spring Cloud Plus，不僅完全覆蓋了 Spring Cloud 的原
生特性，而且提供了更為穩定和成熟的實現。其與 Spring Cloud 的特性比
較如表 8.1 所示。

表 8.1 Spring Cloud、Dubbo Spring Cloud 特性比較

功能元件	Spring Cloud	Dubbo Spring Cloud
分散式設定（Distributed configuration）	Git、Zookeeper、Consul、JDBC	Spring Cloud 分散式設定 + Dubbo 設定中心
服務註冊與發現（Service registration and discovery）	Eureka、Zookeeper、Consul	Spring Cloud 原生註冊中心 + Dubbo 原生註冊中心
負載平衡（Load balancing）	Ribbon（隨機、輪詢等演算法）	Dubbo 內建實現（隨機、輪詢等演算法 + 權重等特性）
服務熔斷（Circuit Breakers）	Spring Cloud Hystrix	Spring Cloud Hystrix + Alibaba Sentinel 等
服務呼叫（Service-to-service calls）	OpenFeign、RestTemplate	Spring Cloud 服務呼叫 + Dubbo @Reference
鏈路追蹤（Tracing）	Spring Cloud Sleuth + Zipkin	Zipkin、Opentracing 等

Dubbo Spring Cloud 的主要特性如下所述。

■ 針對介面代理的高性能遠端程序呼叫：提供了高性能的以代理為基礎的遠端呼叫能力，服務以介面為粒度，隱藏了遠端呼叫底層細節。

■ 智慧負載平衡：內建多種負載平衡策略，可智慧感知下游節點的健康狀況，顯著減少呼叫延遲，加強系統傳輸量。

■ 服務自動註冊與發現：支援多種註冊中心服務，服務實例上下線即時感知。

■ 高度可擴充能力：遵循「微核心 + 外掛程式」的設計原則，所有核心能力如 Protocol、Transport、Serialization 被設計為擴充點，平等對待內建實現和協力廠商實現。

■ 運行期流量排程，內建條件、指令稿等路由策略：透過設定不同的路由規則，可輕鬆實現灰階發佈及同機房優先等功能。

■ 視覺化的服務治理與運行維護：提供豐富的服務治理、運行維護工具，隨時查詢服務中繼資料、服務健康狀態及呼叫統計，即時下發路由策略、調整設定參數。

8.6.2 Kotlin 整合 Spring Cloud Dubbo

定義一個 Maven 子專案：chapter08-dubbo，這是一個 Dubbo 服務提供方。pom 檔案如下：

```xml
1.  <?xml version="1.0" encoding="UTF-8"?>
2.  <project xmlns="http://maven.apache.org/POM/4.0.0"
3.          xmlns:xsi="http://www.w3.org/2001/XMLSchema-instance"
4.          xsi:schemaLocation="http://maven.apache.org/POM/4.0.0 http://
    maven.apache.org/xsd/maven-4.0.0.xsd">
5.      <parent>
6.          <artifactId>kotlinspringboot</artifactId>
7.          <groupId>io.kang.kotlinspringboot</groupId>
8.          <version>0.0.1-SNAPSHOT</version>
9.      </parent>
10.     <modelVersion>4.0.0</modelVersion>
11.     <!-- 子專案名 -->
12.     <artifactId>chapter08-dubbo</artifactId>
13.     <dependencies>
14.         <!-- Spring Boot Actuator 相依套件 -->
15.         <dependency>
16.             <groupId>org.springframework.boot</groupId>
17.             <artifactId>spring-boot-actuator</artifactId>
18.         </dependency>
19.         <!-- Spring Boot Web 相依套件 -->
20.         <dependency>
21.             <groupId>org.springframework.boot</groupId>
22.             <artifactId>spring-boot-starter-web</artifactId>
23.         </dependency>
```

```
24.          <!-- Spring Cloud Dubbo 相依套件 -->
25.      <dependency>
26.          <groupId>com.alibaba.cloud</groupId>
27.          <artifactId>spring-cloud-starter-dubbo</artifactId>
28.          <version>2.2.0.RELEASE</version>
29.      </dependency>
30.          <!-- Spring Cloud Nacos 相依套件 -->
31.      <dependency>
32.          <groupId>com.alibaba.cloud</groupId>
33.          <artifactId>spring-cloud-starter-alibaba-nacos-discovery
    </artifactId>
34.          <version>2.2.0.RELEASE</version>
35.      </dependency>
36.      <dependency>
37.          <groupId>com.fasterxml.jackson.module</groupId>
38.          <artifactId>jackson-module-kotlin</artifactId>
39.      </dependency>
40.      <dependency>
41.          <groupId>org.jetbrains.kotlin</groupId>
42.          <artifactId>kotlin-reflect</artifactId>
43.      </dependency>
44.      <dependency>
45.          <groupId>org.jetbrains.kotlin</groupId>
46.          <artifactId>kotlin-stdlib-jdk8</artifactId>
47.      </dependency>
48.      <dependency>
49.          <groupId>org.jetbrains.kotlinx</groupId>
50.          <artifactId>kotlinx-coroutines-core</artifactId>
51.          <version>1.3.2</version>
52.      </dependency>
53.      <dependency>
54.          <groupId>org.springframework.boot</groupId>
55.          <artifactId>spring-boot-starter-test</artifactId>
56.          <scope>test</scope>
57.      </dependency>
```

```
58.     </dependencies>
59.     <build>
60.         <sourceDirectory>${project.basedir}/src/main/kotlin
    </sourceDirectory>
61.         <testSourceDirectory>${project.basedir}/src/test/kotlin
    </testSourceDirectory>
62.         <plugins>
63.             <plugin>
64.                 <groupId>org.springframework.boot</groupId>
65.                 <artifactId>spring-boot-maven-plugin</artifactId>
66.             </plugin>
67.             <plugin>
68.                 <groupId>org.jetbrains.kotlin</groupId>
69.                 <artifactId>kotlin-maven-plugin</artifactId>
70.                 <configuration>
71.                     <args>
72.                         <arg>-Xjsr305=strict</arg>
73.                     </args>
74.                     <compilerPlugins>
75.                         <plugin>spring</plugin>
76.                         <plugin>jpa</plugin>
77.                     </compilerPlugins>
78.                 </configuration>
79.                 <dependencies>
80.                     <dependency>
81.                         <groupId>org.jetbrains.kotlin</groupId>
82.                         <artifactId>kotlin-maven-allopen</artifactId>
83.                         <version>${kotlin.version}</version>
84.                     </dependency>
85.                     <dependency>
86.                         <groupId>org.jetbrains.kotlin</groupId>
87.                         <artifactId>kotlin-maven-noarg</artifactId>
88.                         <version>${kotlin.version}</version>
89.                     </dependency>
90.                 </dependencies>
```

```
91.            </plugin>
92.          </plugins>
93.      </build>
94. </project>
```

application.yml 檔案的定義如下，定義了 Dubbo 介面所在的套件路徑，使用的協定、通訊埠及 Nacos 註冊中心位址：

```
1.  server:
2.    port: 8087   # 應用通訊埠編號
3.  dubbo:
4.    scan:
5.      # Dubbo 服務掃描基準套件
6.      base-packages: io.kang.provider.dubbo
7.    protocol:
8.      # Dubbo 協定
9.      name: dubbo
10.     # Dubbo 協定通訊埠（ -1 表示自動增加通訊埠，從 20880 開始）
11.     port: -1
12.   registry:
13.     # 掛載到 Nacos 註冊中心
14.     address: nacos://127.0.0.1:8848
15.   cloud:
16.     subscribed-services: ""            # 訂閱的 Dubbo 服務
17. spring:
18.   cloud:
19.     nacos:
20.       discovery:
21.         server-addr: 127.0.0.1:8848     #Nacos 註冊中心位址
22.   application:
23.     name: dubbo-provider                # 應用名稱
```

DubboProviderApp.kt 是啟動類別，開啟了服務註冊註釋：

```
1.  @SpringBootApplication
2.  @EnableDiscoveryClient
```

```
3.   class DubboProviderApp
4.   // 啟動函數
5.   fun main(args: Array<String>) {
6.       runApplication<DubboProviderApp>(*args)
7.   }
```

DubboEchoService.kt 定義了一個對外提供的 Dubbo 服務介面：

```
1.   // 服務介面
2.   interface DubboEchoService {
3.       fun echo(name: String): String
4.   }
```

DubboEchoServiceImpl.kt 實現了 DubboEchoService 介面：

```
1.   // 實現介面
2.   @Service
3.   class DubboEchoServiceImpl: DubboEchoService {
4.       override fun echo(name: String): String {
5.           return "DubboEchoServiceImpl#echo hi $name"
6.       }
7.   }
```

新增一個 Maven 子專案：chapter08-dubbo-consumer，這是一個服務消費方。pom 檔案如下：

```
1.   <?xml version="1.0" encoding="UTF-8"?>
2.   <project xmlns="http://maven.apache.org/POM/4.0.0"
3.           xmlns:xsi="http://www.w3.org/2001/XMLSchema-instance"
4.           xsi:schemaLocation="http://maven.apache.org/POM/4.0.0 http://
     maven.apache.org/xsd/maven-4.0.0.xsd">
5.       <parent>
6.           <artifactId>kotlinspringboot</artifactId>
7.           <groupId>io.kang.kotlinspringboot</groupId>
8.           <version>0.0.1-SNAPSHOT</version>
9.       </parent>
```

```
10.     <modelVersion>4.0.0</modelVersion>
11.     <!-- 子專案名 -->
12.     <artifactId>chapter08-dubbo-consumer</artifactId>
13.     <dependencies>
14.         <!-- 依賴 chapter08-dubbo 子專案 -->
15.         <dependency>
16.             <artifactId>chapter08-dubbo</artifactId>
17.             <groupId>io.kang.kotlinspringboot</groupId>
18.             <version>0.0.1-SNAPSHOT</version>
19.         </dependency>
20.         <!-- Spring Boot Actuator 相依套件 -->
21.         <dependency>
22.             <groupId>org.springframework.boot</groupId>
23.             <artifactId>spring-boot-actuator</artifactId>
24.         </dependency>
25.         <!-- Spring Boot Web 相依套件 -->
26.         <dependency>
27.             <groupId>org.springframework.boot</groupId>
28.             <artifactId>spring-boot-starter-web</artifactId>
29.         </dependency>
30.         <!-- Spring Cloud Dubbo 相依套件 -->
31.         <dependency>
32.             <groupId>com.alibaba.cloud</groupId>
33.             <artifactId>spring-cloud-starter-dubbo</artifactId>
34.             <version>2.2.0.RELEASE</version>
35.         </dependency>
36.         <!-- Spring Cloud Nacos 相依套件 -->
37.         <dependency>
38.             <groupId>com.alibaba.cloud</groupId>
39.             <artifactId>spring-cloud-starter-alibaba-nacos-discovery</artifactId>
40.             <version>2.2.0.RELEASE</version>
41.         </dependency>
42.         <dependency>
43.             <groupId>com.fasterxml.jackson.module</groupId>
```

```
44.            <artifactId>jackson-module-kotlin</artifactId>
45.        </dependency>
46.        <dependency>
47.            <groupId>org.jetbrains.kotlin</groupId>
48.            <artifactId>kotlin-reflect</artifactId>
49.        </dependency>
50.        <dependency>
51.            <groupId>org.jetbrains.kotlin</groupId>
52.            <artifactId>kotlin-stdlib-jdk8</artifactId>
53.        </dependency>
54.        <dependency>
55.            <groupId>org.jetbrains.kotlinx</groupId>
56.            <artifactId>kotlinx-coroutines-core</artifactId>
57.            <version>1.3.2</version>
58.        </dependency>
59.        <dependency>
60.            <groupId>org.springframework.boot</groupId>
61.            <artifactId>spring-boot-starter-test</artifactId>
62.            <scope>test</scope>
63.        </dependency>
64.    </dependencies>
65.    <build>
66.        <sourceDirectory>${project.basedir}/src/main/kotlin
   </sourceDirectory>
67.        <testSourceDirectory>${project.basedir}/src/test/kotlin
   </testSourceDirectory>
68.        <plugins>
69.            <plugin>
70.                <groupId>org.springframework.boot</groupId>
71.                <artifactId>spring-boot-maven-plugin</artifactId>
72.            </plugin>
73.            <plugin>
74.                <groupId>org.jetbrains.kotlin</groupId>
75.                <artifactId>kotlin-maven-plugin</artifactId>
76.                <configuration>
```

```
77.              <args>
78.                  <arg>-Xjsr305=strict</arg>
79.              </args>
80.              <compilerPlugins>
81.                  <plugin>spring</plugin>
82.                  <plugin>jpa</plugin>
83.              </compilerPlugins>
84.          </configuration>
85.          <dependencies>
86.              <dependency>
87.                  <groupId>org.jetbrains.kotlin</groupId>
88.                  <artifactId>kotlin-maven-allopen</artifactId>
89.                  <version>${kotlin.version}</version>
90.              </dependency>
91.              <dependency>
92.                  <groupId>org.jetbrains.kotlin</groupId>
93.                  <artifactId>kotlin-maven-noarg</artifactId>
94.                  <version>${kotlin.version}</version>
95.              </dependency>
96.          </dependencies>
97.      </plugin>
98.    </plugins>
99.  </build>
100. </project>
```

application.yml 檔案中定義了 Nacos 註冊中心位址和訂閱的 Dubbo 服務：

```
1. server:
2.   port: 8088                        # 服務通訊埠編號
3. dubbo:
4.   registry:
5.     # 掛載到 Spring Cloud 註冊中心
6.     address: nacos://127.0.0.1:8848
7.   cloud:
8.     subscribed-services: dubbo-provider   # 訂閱 dubbo-provider 服務
```

```
9.    consumer:
10.     check: false                      # 關閉所有服務的啟動時檢查
11.  spring:
12.    application:
13.      name: dubbo-consumer              # 應用名稱
14.    cloud:
15.      nacos:
16.        discovery:
17.          server-addr: 127.0.0.1:8848  #Nacos 服務註冊中心位址
```

DubboConsumerApp.kt 是啟動類別，開啟了服務註冊註釋：

```
1.  @SpringBootApplication
2.  @EnableDiscoveryClient
3.  class DubboConsumerApp
4.  // 啟動函數
5.  fun main(args: Array<String>) {
6.      runApplication<DubboConsumerApp>(*args)
7.  }
```

CustomerController.kt 定義了一個介面 "/dubbo/echo"，透過 @Reference 註釋存取 Dubbo 服務──dubboEchoService：

```
1.  @RestController
2.  class CustomerController {
3.      // 參考 dubboEchoService
4.      @Reference
5.      lateinit var dubboEchoService: DubboEchoService
6.      // 測試介面，透過 Dubbo 存取 dubboEchoService
7.      @GetMapping("/dubbo/echo/{name}")
8.      fun dubboEcho(@PathVariable("name") name: String): String {
9.          return dubboEchoService.echo(name)
10.      }
11. }
```

8.7 小結

本章介紹了 Spring Cloud Alibaba 生態系統中常用的元件，包含服務註冊和
發現、服務限流降級、訊息驅動、阿里物件雲端儲存、分散式任務排程、
分散式交易、Dubbo 等。通過了解這些元件，你可以在實際的專案中選擇
性地整合並使用它們，以加強開發效率。

Chapter 09

Kotlin 整合服務監控
和服務鏈路監控

微服務系統的分散式特點使得系統監控比較困難，本章將介紹服務監
控中介軟體 Prometheus、Grafana 和服務鏈路監控中介軟體 Zipkin、
SkyWalking，透過範例展示 Kotlin 整合服務監控、服務鏈路監控的方法。

9.1 Prometheus、Grafana 介紹

Prometheus 是由 SoundCloud 開放原始碼的監控警告解決方案。Prometheus
儲存的是時序資料，即按相同時序（相同名稱和標籤），以時間維度儲存
的連續資料的集合。時序（time series）是由名字（metric）以及一組 key/
value 標籤定義的，具有相同名字以及標籤的資料屬於相同時序。metric 有
Counter、Gauge、Histogram、Summary 四種類型。Counter 是一種累加的
metric，如請求的個數、結束的任務數、出現的錯誤數等。Gauge 是正常
的 metric，如溫度，可任意加減，與時間沒有關係，是可以任意變化的資
料。Histogram 是柱狀圖，用於觀察結果取樣、分組及統計。Summary 類
似 Histogram，用於表示一段時間內資料的取樣結果。

Prometheus 提供資料查詢 DSL 語言 ——PromQL。PromQL 支援透過名稱及標籤進行查詢，如 http_requests_total 相等於 {name="http_requests_total"}。查詢準則支援正規比對，如 http_requests_total{code!="200"} 表示查詢 code 不為 "200" 的資料。支援內建函數，如將浮點數轉為整數，檢視每秒資料。支援模糊查詢、比較查詢、範圍查詢、聚合、統計等。查詢結果的類型有：暫態資料（Instant vector），包含一組時序，每個時序只有一個點，舉例來說，http_requests_total；區間資料（Range vector），包含一組時序，每個時序有多個點，舉例來說，http_requests_total[5m]；純量資料（Scalar），純量只有一個數字，沒有時序，舉例來說，count(http_requests_total)。

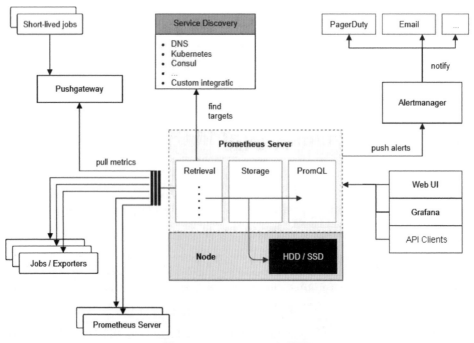

圖 9.1 Prometheus 架構圖

Prometheus 基於 HTTP 採用 pull（拉取）方式收集資料。由於各個被監控的物件不在一個子網或防火牆內導致無法直接拉取各 target 資料，需要將不同資料整理到發送閘道（Pushgateway），再由 Prometheus 統一收集。Prometheus 向 Altermanager 發送警告，架構如圖 9.1 所示。

Grafana 是一款用 Go 語言開發的開放原始碼資料視覺化工具，可用於資料監控和資料統計，帶有警告功能。目前使用 Grafana 的使用者很多。Grafana 具有以下特點。

- **視覺化**：快速且靈活的用戶端具有多種選項，面板外掛程式用不同的方式視覺化指標和記錄檔。
- **警告**：視覺化地為重要的指標定義警示規則，Grafana 將持續評估它們，並發送通知。
- **通知**：警示狀態更改時，Grafana 會發出通知。它還可以接收電子郵件通知。
- **動態儀表板**：使用範本變數建立動態和可重用的儀表板，這些範本變數作為下拉式功能表出現在儀表板頂部。
- **混合資料來源**：在同一張圖中混合不同的資料來源。可以根據每個查詢指定資料來源，這甚至適用於自訂資料來源。
- **註釋**：註釋來自不同的資料來源圖表。將滑鼠指標移過在事件上可以顯示完整的事件中繼資料和標記。
- **篩檢程式**：允許使用者動態建立新的「鍵 - 值」篩檢程式，這些篩檢程式將自動應用於使用該資料來源的所有查詢。

Grafana 從 Prometheus 取得監控指標，並進行視覺化展示，如圖 9.2 所示。

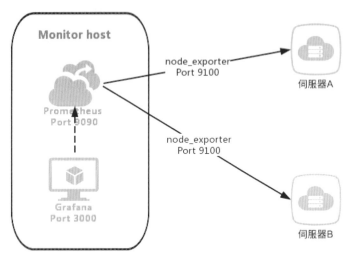

圖 9.2 Prometheus、Grafana 監控服務示意圖

9.2 Kotlin 整合 Prometheus、Grafana

新增一個 Maven 子專案：chapter09-prometheus，pom 檔案如下：

```
1.  <?xml version="1.0" encoding="UTF-8"?>
2.  <project xmlns="http://maven.apache.org/POM/4.0.0"
3.          xmlns:xsi="http://www.w3.org/2001/XMLSchema-instance"
4.          xsi:schemaLocation="http://maven.apache.org/POM/4.0.0 http://
    maven.apache.org/xsd/maven-4.0.0.xsd">
5.      <parent>
6.          <artifactId>kotlinspringboot</artifactId>
7.          <groupId>io.kang.kotlinspringboot</groupId>
8.          <version>0.0.1-SNAPSHOT</version>
9.      </parent>
10.     <modelVersion>4.0.0</modelVersion>
11.     <!-- 子專案名 -->
12.     <artifactId>chapter09-prometheus</artifactId>
13.     <dependencies>
```

```
14.          <!-- Spring Boot Web 相依套件 -->
15.          <dependency>
16.              <groupId>org.springframework.boot</groupId>
17.              <artifactId>spring-boot-starter-web</artifactId>
18.              <version>2.2.1.RELEASE</version>
19.          </dependency>
20.          <!-- Spring Boot Actuator 相依套件 -->
21.          <dependency>
22.              <groupId>org.springframework.boot</groupId>
23.              <artifactId>spring-boot-starter-actuator</artifactId>
24.              <version>2.2.1.RELEASE</version>
25.          </dependency>
26.          <!-- Micrometer 指標相依套件 -->
27.          <dependency>
28.              <groupId>io.micrometer</groupId>
29.              <artifactId>micrometer-registry-prometheus</artifactId>
30.              <version>1.3.5</version>
31.          </dependency>
32.          <dependency>
33.              <groupId>com.fasterxml.jackson.module</groupId>
34.              <artifactId>jackson-module-kotlin</artifactId>
35.          </dependency>
36.          <dependency>
37.              <groupId>org.jetbrains.kotlin</groupId>
38.              <artifactId>kotlin-reflect</artifactId>
39.          </dependency>
40.          <dependency>
41.              <groupId>org.jetbrains.kotlin</groupId>
42.              <artifactId>kotlin-stdlib-jdk8</artifactId>
43.          </dependency>
44.          <dependency>
45.              <groupId>org.jetbrains.kotlinx</groupId>
46.              <artifactId>kotlinx-coroutines-core</artifactId>
47.              <version>1.3.2</version>
48.          </dependency>
```

```
49.    </dependencies>
50.    <build>
51.        <sourceDirectory>${project.basedir}/src/main/kotlin
    </sourceDirectory>
52.        <testSourceDirectory>${project.basedir}/src/test/kotlin
    </testSourceDirectory>
53.        <plugins>
54.            <plugin>
55.                <groupId>org.springframework.boot</groupId>
56.                <artifactId>spring-boot-maven-plugin</artifactId>
57.            </plugin>
58.            <plugin>
59.                <groupId>org.jetbrains.kotlin</groupId>
60.                <artifactId>kotlin-maven-plugin</artifactId>
61.                <configuration>
62.                    <args>
63.                        <arg>-Xjsr305=strict</arg>
64.                    </args>
65.                    <compilerPlugins>
66.                        <plugin>spring</plugin>
67.                        <plugin>jpa</plugin>
68.                    </compilerPlugins>
69.                </configuration>
70.                <dependencies>
71.                    <dependency>
72.                        <groupId>org.jetbrains.kotlin</groupId>
73.                        <artifactId>kotlin-maven-allopen</artifactId>
74.                        <version>${kotlin.version}</version>
75.                    </dependency>
76.                    <dependency>
77.                        <groupId>org.jetbrains.kotlin</groupId>
78.                        <artifactId>kotlin-maven-noarg</artifactId>
79.                        <version>${kotlin.version}</version>
80.                    </dependency>
81.                </dependencies>
```

```
82.              </plugin>
83.          </plugins>
84.      </build>
85. </project>
```

application.yml 檔案的定義如下，曝露了監控介面，Prometheus 透過這些介面擷取監控資訊：

```
1.   server:
2.     port: 8090                    # 應用通訊埠編號
3.   spring:
4.     application:
5.       name: springboot-prometheus # 應用名稱
6.   management:
7.     endpoints:
8.       web:
9.         exposure:
10.          include: '*'            # 對外曝露的監控介面
11.     endpoint:
12.       prometheus:
13.         enabled: true            # 開啟 Prometheus 監控
14.     metrics:
15.       export:
16.         prometheus:
17.           enabled: true          # 輸出 Prometheus 監控指標
18.       tags:
19.         application: ${spring.application.name}  # 監控的標籤
```

PrometheusApp.kt 定義了啟動類別，configurer 方法對外提供監控資訊，這些監控資訊使用 spring.application.name 作為公共標籤：

```
1.   @SpringBootApplication
2.   class PrometheusApp {
3.       // 設定 Micrometer 監控
4.       @Bean
5.       fun configurer(@Value("\${spring.application.name}") applicationName:
```

```
      String):  MeterRegistryCustomizer<MeterRegistry> {
6.        return MeterRegistryCustomizer<MeterRegistry>
7.        { registry -> registry.config().commonTags("application",
   applicationName) }
8.    }
9. }
10. // 啟動函數
11. fun main(args: Array<String>) {
12.    runApplication<PrometheusApp>(*args)
13. }
```

IndexController.kt 提供了一個測試介面，新增一個指標，監控介面存取次數：

```
1. @RestController
2. class IndexController {
3.     private var counter: Counter? = null
4.     constructor(registry: MeterRegistry, @Value("\${spring.application.
   name}") applicationName: String) {
5.        this.counter = registry.counter("index.api.counter",
   "application", applicationName)
6.     }
7.     // 測試介面，每呼叫 1 次，index.api.counter 指標會加 1
8.     @GetMapping("/prometheus")
9.     fun testPrometheus(): String {
10.        this.counter?.increment()
11.        return "hello prometheus"
12.     }
13. }
```

啟動 chapter09-prometheus 專案，並啟動 Prometheus 和 Grafana，可以看到圖 9.3 和圖 9.4 所示的監控頁面。採用 Micrometer 範本，可顯示應用啟動的時間、應用持續的時間及堆積使用率等。

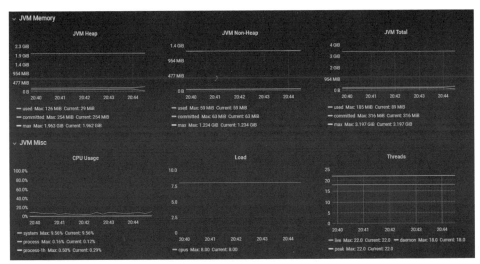

圖 9.3 JVM Memory 指標

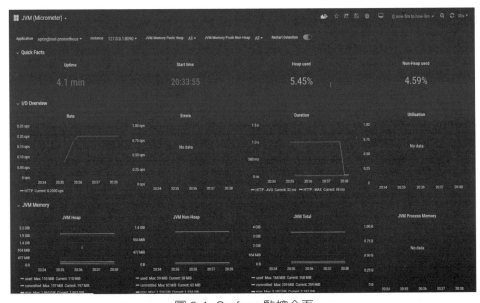

圖 9.4 Grafana 監控介面

此外，還可以自訂監控指標，並顯示在 Grafana 上。舉例來說，對介面 "/prometheus" 的監控，以及監控介面的存取次數，如圖 9.5 所示。

圖 9.5　Grafana 監控自訂指標

9.3 Kotlin 整合 Zipkin

Zipkin 是 Twitter 以 Google 為基礎的分散式監控系統 Dapper（論文）的開放原始碼工具，Zipkin 用於追蹤分散式服務之間的應用資料連結、分析處理延遲時間，可幫助我們改進系統的效能和定位故障。Span 是 Zipkin 的基本工作單元，一次鏈路呼叫就會建立一個 Span。Trace 是一組 Span 的集合，表示一條呼叫鏈路。舉例來說，服務 A 呼叫服務 B 然後呼叫服務 C，這個 A → B → C 的鏈路就是一條 Trace，而每個服務，舉例來說，B 就是一個 Span，如果在服務 B 中另起兩個執行緒分別呼叫了 D、E，那麼 D、E 就是 B 的子 Span。Zipkin 的架構如圖 9.6 所示。

左上部分代表了用戶端，分別為：Instrumented client，是使用了 Zipkin 用戶端工具的服務呼叫方；Instrumented server，是使用了 Zipkin 用戶端工具的服務提供方；Non-Instrumented server，是未使用 Trace 工具的服務提供方，當然還可能存在未使用工具的呼叫方。一個呼叫鏈路是貫穿

Instrumented client → Instrumented server 的，每經過一個服務都會以 Span 的形式透過 Transport 把經過本身的請求上報到 Zipkin 服務端中。

右邊虛線框中的內容代表了 Zipkin 的服務端。UI 提供 Web 頁面，用來展示 Zipkin 中的呼叫鏈和系統相依關係等；Collector 對各個用戶端曝露，負責接收呼叫資料，支援 HTTP、MQ 等；Storage 負責與各個儲存轉換後儲存資料，支援記憶體、MySQL、ES 等；API 為 Web 介面提供查詢儲存中的資料的介面。

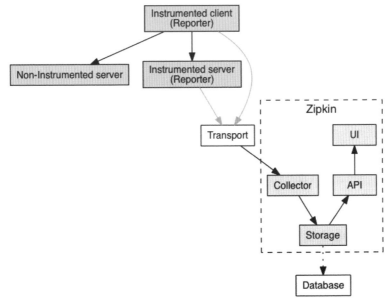

圖 9.6　Zipkin 的架構圖

新增 Maven 子專案 chapter10-zipkin1，這是一個服務提供方，pom 檔案如下：

```
1.  <?xml version="1.0" encoding="UTF-8"?>
2.  <project xmlns="http://maven.apache.org/POM/4.0.0"
3.          xmlns:xsi="http://www.w3.org/2001/XMLSchema-instance"
4.          xsi:schemaLocation="http://maven.apache.org/POM/4.0.0 http://
```

```
     maven.apache.org/xsd/maven-4.0.0.xsd">
5.       <parent>
6.           <artifactId>kotlinspringboot</artifactId>
7.           <groupId>io.kang.kotlinspringboot</groupId>
8.           <version>0.0.1-SNAPSHOT</version>
9.       </parent>
10.      <modelVersion>4.0.0</modelVersion>
11.      <!-- 子專案名 -->
12.      <artifactId>chapter10-zipkin1</artifactId>
13.      <dependencies>
14.          <!-- Spring Boot Web 相依套件 -->
15.          <dependency>
16.              <groupId>org.springframework.boot</groupId>
17.              <artifactId>spring-boot-starter-web</artifactId>
18.              <version>2.2.1.RELEASE</version>
19.          </dependency>
20.          <!-- Spring Cloud Eureka 用戶端相依套件 -->
21.          <dependency>
22.              <groupId>org.springframework.cloud</groupId>
23.              <artifactId>spring-cloud-starter-netflix-eureka-client
     </artifactId>
24.              <version>2.2.1.RELEASE</version>
25.          </dependency>
26.          <!-- Spring Cloud Feign 相依套件 -->
27.          <dependency>
28.              <groupId>org.springframework.cloud</groupId>
29.              <artifactId>spring-cloud-starter-openfeign</artifactId>
30.              <version>2.2.1.RELEASE</version>
31.          </dependency>
32.          <!-- Spring Cloud Zipkin 相依套件 -->
33.          <dependency>
34.              <groupId>org.springframework.cloud</groupId>
35.              <artifactId>spring-cloud-starter-zipkin</artifactId>
36.              <version>2.2.1.RELEASE</version>
37.          </dependency>
```

```xml
38.        <dependency>
39.            <groupId>com.fasterxml.jackson.module</groupId>
40.            <artifactId>jackson-module-kotlin</artifactId>
41.        </dependency>
42.        <dependency>
43.            <groupId>org.jetbrains.kotlin</groupId>
44.            <artifactId>kotlin-reflect</artifactId>
45.        </dependency>
46.        <dependency>
47.            <groupId>org.jetbrains.kotlin</groupId>
48.            <artifactId>kotlin-stdlib-jdk8</artifactId>
49.        </dependency>
50.        <dependency>
51.            <groupId>org.jetbrains.kotlinx</groupId>
52.            <artifactId>kotlinx-coroutines-core</artifactId>
53.            <version>1.3.2</version>
54.        </dependency>
55.    </dependencies>
56.    <build>
57.        <sourceDirectory>${project.basedir}/src/main/kotlin</sourceDirectory>
58.        <testSourceDirectory>${project.basedir}/src/test/kotlin</testSourceDirectory>
59.        <plugins>
60.            <plugin>
61.                <groupId>org.springframework.boot</groupId>
62.                <artifactId>spring-boot-maven-plugin</artifactId>
63.            </plugin>
64.            <plugin>
65.                <groupId>org.jetbrains.kotlin</groupId>
66.                <artifactId>kotlin-maven-plugin</artifactId>
67.                <configuration>
68.                    <args>
69.                        <arg>-Xjsr305=strict</arg>
70.                    </args>
```

```
71.                    <compilerPlugins>
72.                        <plugin>spring</plugin>
73.                        <plugin>jpa</plugin>
74.                    </compilerPlugins>
75.                </configuration>
76.                <dependencies>
77.                    <dependency>
78.                        <groupId>org.jetbrains.kotlin</groupId>
79.                        <artifactId>kotlin-maven-allopen</artifactId>
80.                        <version>${kotlin.version}</version>
81.                    </dependency>
82.                    <dependency>
83.                        <groupId>org.jetbrains.kotlin</groupId>
84.                        <artifactId>kotlin-maven-noarg</artifactId>
85.                        <version>${kotlin.version}</version>
86.                    </dependency>
87.                </dependencies>
88.            </plugin>
89.        </plugins>
90.    </build>
91. </project>
```

application.yml 檔案的內容如下：

```
1.  server:
2.    port: 8102                                      # 應用通訊埠編號
3.  spring:
4.    application:
5.      name: zipkin-service2                         # 應用名
6.    zipkin:
7.      base-url: http://localhost:9411               #Zipkin 伺服器位址
8.  eureka:
9.    client:
10.     service-url:
11.       defaultZone: http://localhost:8761/eureka/  # 服務註冊中心位址
```

ZipkinApplication1.kt 啟動了一個 Spring Boot 應用：

```
1.  // 開啟 Eureka Client 註釋
2.  @SpringBootApplication
3.  @EnableEurekaClient
4.  class ZipkinApplication1 {
5.      // 開啟負載平衡
6.      @Bean
7.      @LoadBalanced
8.      fun restTemplate(): RestTemplate {
9.          return RestTemplate()
10.     }
11. }
12. // 啟動函數
13. fun main(args: Array<String>) {
14.     runApplication<ZipkinApplication1>(*args)
15. }
```

ZipkinController.kt 定義了一個測試介面 "service2"：

```
1.  @RestController
2.  class ZipkinController {
3.      // 測試介面
4.      @GetMapping("service2")
5.      fun service2(): String {
6.          return "Hello, I'm service2"
7.      }
8.  }
```

新增 Maven 子專案 chapter10-zipkin，這是一個服務消費方，pom 檔案如下：

```
1.  <?xml version="1.0" encoding="UTF-8"?>
2.  <project xmlns="http://maven.apache.org/POM/4.0.0"
3.          xmlns:xsi="http://www.w3.org/2001/XMLSchema-instance"
4.          xsi:schemaLocation="http://maven.apache.org/POM/4.0.0 http://
```

```
     maven.apache.org/xsd/maven-4.0.0.xsd">
5.       <parent>
6.           <artifactId>kotlinspringboot</artifactId>
7.           <groupId>io.kang.kotlinspringboot</groupId>
8.           <version>0.0.1-SNAPSHOT</version>
9.       </parent>
10.      <modelVersion>4.0.0</modelVersion>
11.      <!-- 子專案名 -->
12.      <artifactId>chapter10-zipkin</artifactId>
13.      <dependencies>
14.          <!-- Spring Boot Web 相依套件 -->
15.          <dependency>
16.              <groupId>org.springframework.boot</groupId>
17.              <artifactId>spring-boot-starter-web</artifactId>
18.              <version>2.2.1.RELEASE</version>
19.          </dependency>
20.          <!-- Spring Cloud Eureka 用戶端相依套件 -->
21.          <dependency>
22.              <groupId>org.springframework.cloud</groupId>
23.              <artifactId>spring-cloud-starter-netflix-eureka-client
     </artifactId>
24.              <version>2.2.1.RELEASE</version>
25.          </dependency>
26.          <!-- Spring Cloud Feign 相依套件 -->
27.          <dependency>
28.              <groupId>org.springframework.cloud</groupId>
29.              <artifactId>spring-cloud-starter-openfeign</artifactId>
30.              <version>2.2.1.RELEASE</version>
31.          </dependency>
32.          <!-- Spring Cloud Zipkin 相依套件 -->
33.          <dependency>
34.              <groupId>org.springframework.cloud</groupId>
35.              <artifactId>spring-cloud-starter-zipkin</artifactId>
36.              <version>2.2.1.RELEASE</version>
37.          </dependency>
```

```
38.        <dependency>
39.            <groupId>com.fasterxml.jackson.module</groupId>
40.            <artifactId>jackson-module-kotlin</artifactId>
41.        </dependency>
42.        <dependency>
43.            <groupId>org.jetbrains.kotlin</groupId>
44.            <artifactId>kotlin-reflect</artifactId>
45.        </dependency>
46.        <dependency>
47.            <groupId>org.jetbrains.kotlin</groupId>
48.            <artifactId>kotlin-stdlib-jdk8</artifactId>
49.        </dependency>
50.        <dependency>
51.            <groupId>org.jetbrains.kotlinx</groupId>
52.            <artifactId>kotlinx-coroutines-core</artifactId>
53.            <version>1.3.2</version>
54.        </dependency>
55.    </dependencies>
56.    <build>
57.        <sourceDirectory>${project.basedir}/src/main/kotlin</sourceDirectory>
58.        <testSourceDirectory>${project.basedir}/src/test/kotlin</testSourceDirectory>
59.        <plugins>
60.            <plugin>
61.                <groupId>org.springframework.boot</groupId>
62.                <artifactId>spring-boot-maven-plugin</artifactId>
63.            </plugin>
64.            <plugin>
65.                <groupId>org.jetbrains.kotlin</groupId>
66.                <artifactId>kotlin-maven-plugin</artifactId>
67.                <configuration>
68.                    <args>
69.                        <arg>-Xjsr305=strict</arg>
70.                    </args>
```

```
71.                    <compilerPlugins>
72.                        <plugin>spring</plugin>
73.                        <plugin>jpa</plugin>
74.                    </compilerPlugins>
75.                </configuration>
76.                <dependencies>
77.                    <dependency>
78.                        <groupId>org.jetbrains.kotlin</groupId>
79.                        <artifactId>kotlin-maven-allopen</artifactId>
80.                        <version>${kotlin.version}</version>
81.                    </dependency>
82.                    <dependency>
83.                        <groupId>org.jetbrains.kotlin</groupId>
84.                        <artifactId>kotlin-maven-noarg</artifactId>
85.                        <version>${kotlin.version}</version>
86.                    </dependency>
87.                </dependencies>
88.            </plugin>
89.        </plugins>
90.    </build>
91. </project>
```

application.yml 檔案的內容如下：

```
1.  server:
2.    port: 8101                              # 應用通訊埠編號
3.  spring:
4.    application:
5.      name: zipkin-service1                 # 應用名
6.    zipkin:
7.      base-url: http://localhost:9411       # Zipkin 服務位址
8.  eureka:
9.    client:
10.     service-url:
11.       defaultZone: http://localhost:8761/eureka/  # Eureka 註冊中心位址
```

ZipkinApplication.kt 的定義如下，這是一個啟動類別，啟動了一個 Spring Boot 服務：

```
1.  // 開啟 Eureka Client 註釋
2.  @SpringBootApplication
3.  @EnableEurekaClient
4.  class ZipkinApplication {
5.      @Bean
6.      @LoadBalanced
7.      fun restTemplate(): RestTemplate {
8.          return RestTemplate()
9.      }
10. }
11. // 啟動函數
12. fun main(args: Array<String>) {
13.     runApplication<ZipkinApplication>(*args)
14. }
```

ZipkinController.kt，呼叫 zipkin-service2 提供的服務：

```
1.  @RestController
2.  class ZipkinController {
3.      @Autowired
4.      lateinit var restTemplate: RestTemplate
5.      // 測試介面，使用 Ribbon 方式呼叫 /service2 介面
6.      @GetMapping("service1")
7.      fun service1(): String {
8.          return restTemplate.getForObject("http://zipkin-service2/
    service2", String::class)
9.      }
10. }
```

透過 "service1" 呼叫 "service2" 介面，在 ZipKin 監控介面可以看到呼叫資訊，每次呼叫都有記錄，如圖 9.7 所示。

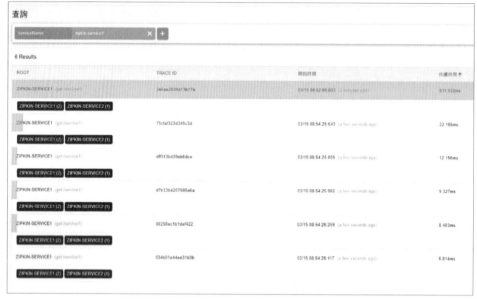

圖 9.7 Zipkin 監控介面

實際到某次呼叫,可以看到呼叫每個服務介面的時間,如圖 9.8 所示。

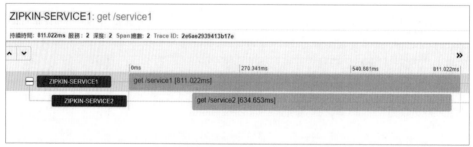

圖 9.8 Zipkin 監控服務介面呼叫時間

此外,還可以展示服務的相依關係,如圖 9.9 所示。

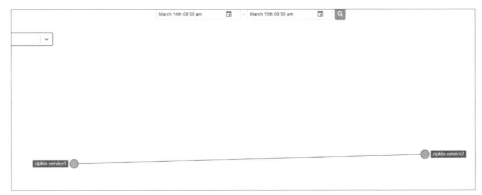

圖 9.9 服務相依關係圖

9.4 Kotlin 整合 SkyWalking

SkyWalking 是由中文社群開放原始碼同好吳晟開放原始碼並提交到
Apache 孵化器的產品，它同時吸收了 Zipkin/Pinpoint/CAT 的設計想法，
支援非侵入式埋點。這是一款以分散式追蹤為基礎的應用程式效能監控系
統。SkyWalking 包含分散式追蹤、效能指標分析、應用和服務依賴分析
等，其架構如圖 9.10 所示。

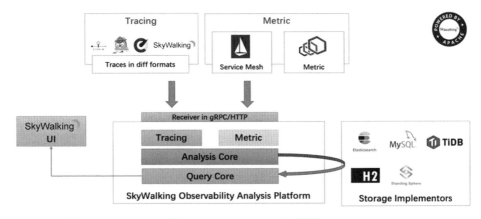

圖 9.10 SkyWalking 架構圖

SkyWalking 的核心是一個資料分析和度量結果的儲存平台，透過 HTTP 或 gRPC 方式向 SkyWalking Collecter 提交分析和度量資料，SkyWalking Collecter 對資料進行分析和聚合，並將資料儲存到 Elasticsearch、H2、MySQL、TiDB 等其一，最後可以透過 SkyWalking UI 的視覺化介面對最後的結果進行呈現。SkyWalking 支援從多個來源和多種格式收集資料——多種語言的 SkyWalking Agent、Zipkin v1/v2、Istio 勘測、Envoy 度量等資料格式。

SkyWalking 收集各種格式的資料進行儲存，然後進行展示。架設 SkyWalking 服務需要關注的是 SkyWalking Collecter、SkyWalking UI 和存放裝置。

SkyWalking 採用 Java 探針技術，在程式層面對應用程式沒有任何侵入，使用起來簡單方便，當然其實作方式需要針對不同的架構及服務提供探針外掛程式。

新增 Maven 子專案 chapter10-skywalking1，這是一個服務提供方，pom 檔案如下：

```
1.  <?xml version="1.0" encoding="UTF-8"?>
2.  <project xmlns="http://maven.apache.org/POM/4.0.0"
3.          xmlns:xsi="http://www.w3.org/2001/XMLSchema-instance"
4.          xsi:schemaLocation="http://maven.apache.org/POM/4.0.0 http://maven.apache.org/xsd/maven-4.0.0.xsd">
5.      <parent>
6.          <artifactId>kotlinspringboot</artifactId>
7.          <groupId>io.kang.kotlinspringboot</groupId>
8.          <version>0.0.1-SNAPSHOT</version>
9.      </parent>
10.     <modelVersion>4.0.0</modelVersion>
11.     <!-- 子專案名 -->
12.     <artifactId>chapter10-skywalking1</artifactId>
13.     <dependencies>
```

```
14.          <!-- Spring Boot Web 相依套件 -->
15.      <dependency>
16.          <groupId>org.springframework.boot</groupId>
17.          <artifactId>spring-boot-starter-web</artifactId>
18.          <version>2.2.1.RELEASE</version>
19.      </dependency>
20.          <!-- Spring Boot Nacos 相依套件 -->
21.      <dependency>
22.          <groupId>com.alibaba.cloud</groupId>
23.          <artifactId>spring-cloud-starter-alibaba-nacos-discovery
   </artifactId>
24.          <version>2.1.1.RELEASE</version>
25.      </dependency>
26.          <!-- Spring Boot Feign 相依套件 ->
27.      <dependency>
28.          <groupId>org.springframework.cloud</groupId>
29.          <artifactId>spring-cloud-starter-openfeign</artifactId>
30.          <version>2.2.1.RELEASE</version>
31.      </dependency>
32.      <dependency>
33.          <groupId>com.fasterxml.jackson.module</groupId>
34.          <artifactId>jackson-module-kotlin</artifactId>
35.      </dependency>
36.      <dependency>
37.          <groupId>org.jetbrains.kotlin</groupId>
38.          <artifactId>kotlin-reflect</artifactId>
39.      </dependency>
40.      <dependency>
41.          <groupId>org.jetbrains.kotlin</groupId>
42.          <artifactId>kotlin-stdlib-jdk8</artifactId>
43.      </dependency>
44.      <dependency>
45.          <groupId>org.jetbrains.kotlinx</groupId>
46.          <artifactId>kotlinx-coroutines-core</artifactId>
47.          <version>1.3.2</version>
48.      </dependency>
```

```
49.    </dependencies>
50.    <build>
51.        <sourceDirectory>${project.basedir}/src/main/kotlin
    </sourceDirectory>
52.        <testSourceDirectory>${project.basedir}/src/test/kotlin
    </testSourceDirectory>
53.        <plugins>
54.            <plugin>
55.                <groupId>org.springframework.boot</groupId>
56.                <artifactId>spring-boot-maven-plugin</artifactId>
57.            </plugin>
58.            <plugin>
59.                <groupId>org.jetbrains.kotlin</groupId>
60.                <artifactId>kotlin-maven-plugin</artifactId>
61.                <configuration>
62.                    <args>
63.                        <arg>-Xjsr305=strict</arg>
64.                    </args>
65.                    <compilerPlugins>
66.                        <plugin>spring</plugin>
67.                        <plugin>jpa</plugin>
68.                    </compilerPlugins>
69.                </configuration>
70.                <dependencies>
71.                    <dependency>
72.                        <groupId>org.jetbrains.kotlin</groupId>
73.                        <artifactId>kotlin-maven-allopen</artifactId>
74.                        <version>${kotlin.version}</version>
75.                    </dependency>
76.                    <dependency>
77.                        <groupId>org.jetbrains.kotlin</groupId>
78.                        <artifactId>kotlin-maven-noarg</artifactId>
79.                        <version>${kotlin.version}</version>
80.                    </dependency>
81.                </dependencies>
82.            </plugin>
```

```
83.        </plugins>
84.      </build>
85. </project>
```

application.yml 的定義如下，使用 Nacos 做服務註冊中心：

```
1.  server:
2.    port: 8103                        # 應用通訊埠編號
3.  spring:
4.    application:
5.      name: skywalking-service2       # 應用名
6.    cloud:
7.      nacos:
8.        discovery:
9.          server-addr: 127.0.0.1:8848    #Nacos 服務註冊中心位址
```

SkyWalkingApp1.kt 啟動一個 Spring Boot 應用：

```
1.  // 開啟服務發現註釋
2.  @SpringBootApplication
3.  @EnableDiscoveryClient
4.  class SkyWalkingApp1 {
5.      @Bean
6.      @LoadBalanced
7.      fun restTemplate(): RestTemplate {
8.          return RestTemplate()
9.      }
10. }
11. // 啟動函數
12. fun main(args: Array<String>) {
13.     runApplication<SkyWalkingApp1>(*args)
14. }
```

SkyWalkingController.kt 提供一個服務介面 "service2"：

```
1.  @RestController
2.  class SkyWalkingController {
```

```
3.      // 測試介面
4.      @GetMapping("service2")
5.      fun service2(): String {
6.          return "Hello, I'm SkyWalking"
7.      }
8.  }
```

新增一個 Maven 子專案：chapter10-skywalking，這是一個服務消費者，pom 檔案如下：

```
1.  <?xml version="1.0" encoding="UTF-8"?>
2.  <project xmlns="http://maven.apache.org/POM/4.0.0"
3.          xmlns:xsi="http://www.w3.org/2001/XMLSchema-instance"
4.          xsi:schemaLocation="http://maven.apache.org/POM/4.0.0 http://
    maven.apache.org/xsd/maven-4.0.0.xsd">
5.      <parent>
6.          <artifactId>kotlinspringboot</artifactId>
7.          <groupId>io.kang.kotlinspringboot</groupId>
8.          <version>0.0.1-SNAPSHOT</version>
9.      </parent>
10.     <modelVersion>4.0.0</modelVersion>
11.     <!-- 子專案名 -->
12.     <artifactId>chapter10-skywalking</artifactId>
13.     <dependencies>
14.         <!-- Spring Boot Web 相依套件 -->
15.         <dependency>
16.             <groupId>org.springframework.boot</groupId>
17.             <artifactId>spring-boot-starter-web</artifactId>
18.             <version>2.2.1.RELEASE</version>
19.         </dependency>
20.         <!-- Spring Cloud Nacos 相依套件 -->
21.         <dependency>
22.             <groupId>com.alibaba.cloud</groupId>
23.             <artifactId>spring-cloud-starter-alibaba-nacos-discovery
    </artifactId>
```

```
24.                <version>2.1.1.RELEASE</version>
25.            </dependency>
26.            <!-- Spring Cloud Feign 相依套件 -->
27.            <dependency>
28.                <groupId>org.springframework.cloud</groupId>
29.                <artifactId>spring-cloud-starter-openfeign</artifactId>
30.                <version>2.2.1.RELEASE</version>
31.            </dependency>
32.            <dependency>
33.                <groupId>com.fasterxml.jackson.module</groupId>
34.                <artifactId>jackson-module-kotlin</artifactId>
35.            </dependency>
36.            <dependency>
37.                <groupId>org.jetbrains.kotlin</groupId>
38.                <artifactId>kotlin-reflect</artifactId>
39.            </dependency>
40.            <dependency>
41.                <groupId>org.jetbrains.kotlin</groupId>
42.                <artifactId>kotlin-stdlib-jdk8</artifactId>
43.            </dependency>
44.            <dependency>
45.                <groupId>org.jetbrains.kotlinx</groupId>
46.                <artifactId>kotlinx-coroutines-core</artifactId>
47.                <version>1.3.2</version>
48.            </dependency>
49.        </dependencies>
50.        <build>
51.            <sourceDirectory>${project.basedir}/src/main/kotlin
    </sourceDirectory>
52.            <testSourceDirectory>${project.basedir}/src/test/kotlin
    </testSourceDirectory>
53.            <plugins>
54.                <plugin>
55.                    <groupId>org.springframework.boot</groupId>
56.                    <artifactId>spring-boot-maven-plugin</artifactId>
```

```
57.          </plugin>
58.          <plugin>
59.              <groupId>org.jetbrains.kotlin</groupId>
60.              <artifactId>kotlin-maven-plugin</artifactId>
61.              <configuration>
62.                  <args>
63.                      <arg>-Xjsr305=strict</arg>
64.                  </args>
65.                  <compilerPlugins>
66.                      <plugin>spring</plugin>
67.                      <plugin>jpa</plugin>
68.                  </compilerPlugins>
69.              </configuration>
70.              <dependencies>
71.                  <dependency>
72.                      <groupId>org.jetbrains.kotlin</groupId>
73.                      <artifactId>kotlin-maven-allopen</artifactId>
74.                      <version>${kotlin.version}</version>
75.                  </dependency>
76.                  <dependency>
77.                      <groupId>org.jetbrains.kotlin</groupId>
78.                      <artifactId>kotlin-maven-noarg</artifactId>
79.                      <version>${kotlin.version}</version>
80.                  </dependency>
81.              </dependencies>
82.          </plugin>
83.      </plugins>
84.  </build>
85. </project>
```

application.yml 檔案中的定義如下，使用 Nacos 做服務發現：

```
1. server:
2.   port: 8102                          # 應用通訊埠編號
3. spring:
4.   application:
```

```
5.      name: skywalking-service1        # 應用名
6.    cloud:
7.      nacos:
8.        discovery:
9.          server-addr: 127.0.0.1:8848    # Nacos 服務註冊中心位址
```

SkyWalkingApp.kt 啟動一個 Spring Boot 服務：

```
1.  // 開啟服務註冊註釋
2.  @SpringBootApplication
3.  @EnableDiscoveryClient
4.  class SkyWalkingApp {
5.      @Bean
6.      @LoadBalanced
7.      fun restTemplate(): RestTemplate {
8.          return RestTemplate()
9.      }
10. }
11. // 啟動函數
12. fun main(args: Array<String>) {
13.     runApplication<SkyWalkingApp>(*args)
14. }
```

SkyWalkingController.kt 呼叫 "service2" 介面：

```
1.  @RestController
2.  class SkyWalkingController {
3.      @Autowired
4.      lateinit var restTemplate: RestTemplate
5.      // 測試介面，透過 Ribbon 方式呼叫 service2 介面
6.      @GetMapping("service1")
7.      fun service1(): String {
8.          return restTemplate.getForObject("http://skywalking-service2/
    service2", String::class)
9.      }
10. }
```

分別在 VM Options 填寫以下參數,然後執行 SkyWalkingApp1.kt、
SkyWalkingApp.kt,啟動服務提供方和服務消費方,透過代理方式,對程
式無侵入,可監控服務間的呼叫。呼叫 "service1" 介面幾次:

```
1.  -javaagent:D:\soft\apache-skywalking-apm-6.6.0\apache-skywalking-apm-bin\
    agent\skywalking-agent.jar
2.  -Dskywalking.agent.service_name=skywalking-service1
3.  -Dskywalking.collector.backend_service=localhost:11800
4.
5.  -javaagent:D:\soft\apache-skywalking-apm-6.6.0\apache-skywalking-apm-bin\
    agent\skywalking-agent.jar
6.  -Dskywalking.agent.service_name=skywalking-service2
7.  -Dskywalking.collector.backend_service=localhost:11800
```

圖 9.11 SkyWalking 監控介面

在 SkyWalking 監控介面可以看到有兩個 service，分別是 skywalking-service1 和 skywalking-service2，和 VM Options 參數定義的一致，如圖 9.11 所示。

可以在圖 9.12 中看到每次呼叫經過哪些介面和對應的耗時。

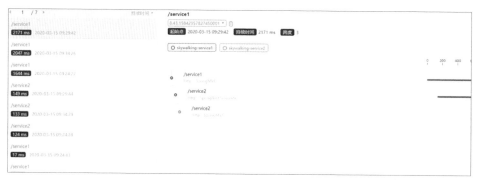

圖 9.12　SkyWalking 監控介面呼叫耗時

在圖 9.13 中可以看到服務的拓撲關係。

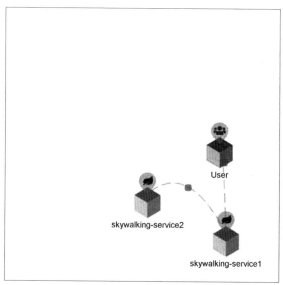

圖 9.13　服務拓撲關係

9.5 小結

本章介紹了使用 Kotlin 整合 Prometheus 和 Grafana 進行服務監控的方法。
Prometheus 可以定時擷取服務效能指標，Grafana 提供監控範本和優美的
監控介面。本書使用 Micrometer 範本，提供 JVM 各種監控指標，此外還
定義了一個介面呼叫次數監控指標。

本章還介紹了服務監控工具 ZipKin 和 Skywalking，它們有助監控系統的
呼叫情況，找到系統性能瓶頸，並有針對性地加強系統性能。

Chapter 10

基於 Kotlin 和 Spring Boot 架設部落格

本章將在之前章節的基礎上，透過一個完整的專案，介紹如何使用 Kotlin 和 Spring Boot 開發微服務系統。本章以一個部落格系統為例，介紹初始化 Maven 專案、系統整體架構、資料物理定義、資料庫設計、資料庫操作層、Service 層及 Controller 層的相關知識。最後，介紹如何把該專案部署到騰訊雲，並展示部落格的效果。透過這個專案，讀者可以架設屬於自己的部落格平台，記錄生活中的點滴。

10.1 初始化 Maven 專案

基於 Kotlin 和 Spring Boot 架設部落格，會用到表 10.1 列出的各項技術。

表 10.1　部落格使用的技術堆疊列表

技術堆疊	介紹
Thymeleaf	範本引擎
SpringBoot	Spring 快速開發架構
Spring Session	session 管理

技術堆疊	介紹
Kaptcha	驗證碼
CommonMark	Markdown 工具
MySQL	資料庫
Spring Data Jpa	持久化層架構
QueryDSL	通用查詢架構
Kotlin	開發語言

新增 Maven 子專案 chapter11-blog，pom.xml 如下：

```
1.  <?xml version="1.0" encoding="UTF-8"?>
2.  <project xmlns="http://maven.apache.org/POM/4.0.0"
3.          xmlns:xsi="http://www.w3.org/2001/XMLSchema-instance"
4.          xsi:schemaLocation="http://maven.apache.org/POM/4.0.0 http://
    maven.apache.org/xsd/maven-4.0.0.xsd">
5.      <parent>
6.          <artifactId>kotlinspringboot</artifactId>
7.          <groupId>io.kang.kotlinspringboot</groupId>
8.          <version>0.0.1-SNAPSHOT</version>
9.      </parent>
10.     <modelVersion>4.0.0</modelVersion>
11.     <!-- 子專案名 -->
12.     <artifactId>chapter11-blog</artifactId>
13.     <dependencies>
14.         <!-- Spring Boot Web 相依套件 -->
15.         <dependency>
16.             <groupId>org.springframework.boot</groupId>
17.             <artifactId>spring-boot-starter-web</artifactId>
18.         </dependency>
19.         <!-- Spring Boot Thymeleaf 範本引擎相依套件 -->
20.         <dependency>
21.             <groupId>org.springframework.boot</groupId>
22.             <artifactId>spring-boot-starter-thymeleaf</artifactId>
```

```
23.        </dependency>
24.         <!-- Spring Session 相依套件 -->
25.        <dependency>
26.            <groupId>org.springframework.session</groupId>
27.            <artifactId>spring-session-core</artifactId>
28.        </dependency>
29.         <!-- 驗證碼 -->
30.        <dependency>
31.            <groupId>com.github.penggle</groupId>
32.            <artifactId>kaptcha</artifactId>
33.            <version>2.3.2</version>
34.        </dependency>
35.         <!-- CommonMark Core 相依套件 -->
36.        <dependency>
37.            <groupId>com.atlassian.commonmark</groupId>
38.            <artifactId>commonmark</artifactId>
39.            <version>0.8.0</version>
40.        </dependency>
41.         <!-- CommonMark Table 相依套件 -->
42.        <dependency>
43.            <groupId>com.atlassian.commonmark</groupId>
44.            <artifactId>commonmark-ext-gfm-tables</artifactId>
45.            <version>0.8.0</version>
46.        </dependency>
47.         <!-- MySQL 驅動相依套件 -->
48.        <dependency>
49.            <groupId>mysql</groupId>
50.            <artifactId>mysql-connector-java</artifactId>
51.            <scope>runtime</scope>
52.        </dependency>
53.         <!-- Spring Data JPA 相依套件 -->
54.        <dependency>
55.            <groupId>org.springframework.boot</groupId>
56.            <artifactId>spring-boot-starter-data-jpa</artifactId>
57.        </dependency>
```

```
58.         <!-- QueryDSL 相依套件 -->
59.     <dependency>
60.         <groupId>com.querydsl</groupId>
61.         <artifactId>querydsl-jpa</artifactId>
62.     </dependency>
63.     <dependency>
64.         <groupId>com.querydsl</groupId>
65.         <artifactId>querydsl-apt</artifactId>
66.     </dependency>
67.         <!-- Spring Boot Test 相依套件 -->
68.     <dependency>
69.         <groupId>org.springframework.boot</groupId>
70.         <artifactId>spring-boot-starter-test</artifactId>
71.         <scope>test</scope>
72.     </dependency>
73.     <dependency>
74.         <groupId>com.fasterxml.jackson.module</groupId>
75.         <artifactId>jackson-module-kotlin</artifactId>
76.     </dependency>
77.     <dependency>
78.         <groupId>org.jetbrains.kotlin</groupId>
79.         <artifactId>kotlin-reflect</artifactId>
80.     </dependency>
81.     <dependency>
82.         <groupId>org.jetbrains.kotlin</groupId>
83.         <artifactId>kotlin-stdlib-jdk8</artifactId>
84.     </dependency>
85.     <dependency>
86.         <groupId>org.jetbrains.kotlinx</groupId>
87.         <artifactId>kotlinx-coroutines-core</artifactId>
88.         <version>1.3.2</version>
89.     </dependency>
90.     </dependencies>
91.     <build>
92.         <sourceDirectory>${project.basedir}/src/main/kotlin</sourceDirectory>
```

```
93.        <testSourceDirectory>${project.basedir}/src/test/kotlin
   </testSourceDirectory>
94.        <plugins>
95.            <plugin>
96.                <groupId>org.springframework.boot</groupId>
97.                <artifactId>spring-boot-maven-plugin</artifactId>
98.            </plugin>
99.            <plugin>
100.                <groupId>com.querydsl</groupId>
101.                <artifactId>querydsl-maven-plugin</artifactId>
102.                <executions>
103.                    <execution>
104.                        <phase>compile</phase>
105.                        <goals>
106.                            <goal>jpa-export</goal>
107.                        </goals>
108.                        <configuration>
109.                            <targetFolder>target/generated-sources/
   kotlin</targetFolder>
110.                            <packages>io.kang.blog.entity</packages>
111.                        </configuration>
112.                    </execution>
113.                </executions>
114.            </plugin>
115.            <plugin>
116.                <groupId>org.jetbrains.kotlin</groupId>
117.                <artifactId>kotlin-maven-plugin</artifactId>
118.                <configuration>
119.                    <args>
120.                        <arg>-Xjsr305=strict</arg>
121.                    </args>
122.                    <compilerPlugins>
123.                        <plugin>spring</plugin>
124.                        <plugin>jpa</plugin>
125.                    </compilerPlugins>
```

```
126.                         </configuration>
127.                         <dependencies>
128.                             <dependency>
129.                                 <groupId>org.jetbrains.kotlin</groupId>
130.                                 <artifactId>kotlin-maven-allopen</artifactId>
131.                                 <version>${kotlin.version}</version>
132.                             </dependency>
133.                             <dependency>
134.                                 <groupId>org.jetbrains.kotlin</groupId>
135.                                 <artifactId>kotlin-maven-noarg</artifactId>
136.                                 <version>${kotlin.version}</version>
137.                             </dependency>
138.                         </dependencies>
139.                     </plugin>
140.                 </plugins>
141.             </build>
142.         </project>
```

application.yml 檔案的內容如下，其中定義了資料庫連接資訊、資料庫連
接池設定及應用名稱：

```
1.   server:
2.     port: 8111                              # 應用通訊埠編號
3.   spring:
4.     thymeleaf:
5.       cache: false                          #Thymeleaf 不開啟快取
6.     datasource:
7.       driver-class-name: com.mysql.cj.jdbc.Driver   # 資料庫驅動
8.       url: jdbc:mysql://localhost:3306/my_blog_db?useUnicode=
     true&characterEncoding=utf8&autoReconnect=true&useSSL=
     false&serverTimezone=UTC              # 資料庫連接串
9.       username: root                        # 資料庫連接使用者名稱
10.      password: 123456                      # 資料庫連接密碼
11.      hikari:
12.        minimum-idle: 5                     # 連接池空閒連接的最小數量
```

```
13.        maximum-pool-size: 15          # 最大連接池大小
14.        auto-commit: true              # 自動提交
15.        idle-timeout: 30000            # 閒置時間，30s
16.        pool-name: hikariCP            # 連接池名稱
17.        max-lifetime: 1800000          # 一個連接到連接池的存活時間，30 分鐘
18.        connection-timeout: 30000      # 資料庫連接逾時，30s
19.    application:
20.      name: Blog                       # 應用名稱
```

10.2 系統架構

本章介紹的部落格的系統架構如圖 10.1 所示。自下而上分為基礎資源層、物理層、資料操作層、服務層和介面層。部落格依賴 MySQL 資料庫。

物理層分為管理使用者實體、部落格實體、部落格分類實體、部落格評論實體、部落格設定實體，部落格連結實體和部落格標籤實體，每個實體對應資料庫中的一張表，每個實體描述部落格系統的不同維度屬性。

每個實體有對應的資料操作 DAO，本書使用 JPA 和 QueryDSL 實現管理使用者、部落格、部落格分類、部落格評論、部落格設定、部落格連結及部落格標籤的查詢、更新、插入等基礎操作和分頁查詢、批次更新等進階操作。

服務層在物理層和資料操作層的基礎上提供相關的操作。根據功能，服務層劃分為管理使用者服務、部落格服務、部落格分類服務、部落格評論服務、部落格設定服務、部落格連結服務及部落格標籤服務。

介面層提供前台展示介面和後台管理介面。前台展示介面主要提供首頁部落格展示、類別展示、詳情展示、標籤列表展示、評論操作、友情連結展示、關於頁面展示等功能。後台管理介面提供管理使用者相關、部落格編

輯相關、發佈相關、類別管理相關、評論審核相關、設定相關、友情連結
相關、標籤相關及上傳相關介面。

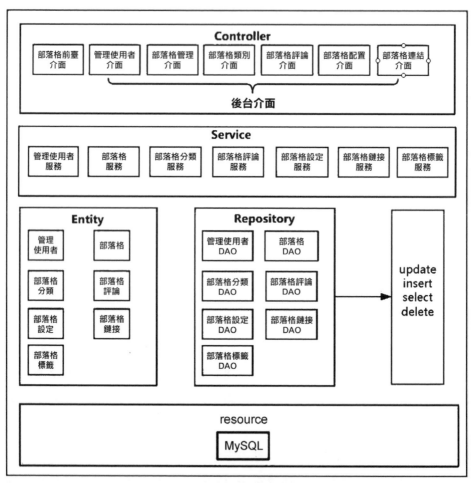

圖 10.1 部落格系統架構圖

10.3 定義實體

部落格系統定義了表 10.2 所示的實體。

表 10.2 部落格定義的實體清單

序號	實體	說明	屬性
1	AdminUser	管理使用者	使用者 id、登入使用者名稱、登入密碼、暱稱、是否鎖定
2	Blog	部落格	部落格 id、標題、url、封面圖片、分類 id、分類名稱、標籤 id、部落格審核狀態、點擊量、是否允許評論、是否刪除、建立時間、更新時間、部落格內容
3	BlogCategory	部落格分類	分類 id、分類名稱、圖示、排名、是否刪除、建立時間
4	BlogComment	部落格評論	評論 id、部落格 id、評論者、電子郵件、網址、評論內容、建立時間、評論者 IP 位址、回覆內容、回覆時間、評論審核狀態、是否刪除
5	BlogConfig	部落格設定	設定名稱、設定值、建立時間、更新時間
6	BlogLink	部落格連結	友情連結 id、類型、名稱、url、描述、排名、是否刪除、建立時間
7	BlogTag	部落格標籤	標籤 id、名稱、是否刪除、建立時間
8	BlogTagRelation	部落格標籤關係	關係 id、部落格 id、標籤 id、建立時間

AdminUser.kt 定義了管理使用者實體，包含 adminUserId、loginUserName、loginPassword、nickName、locked 這幾個屬性。對應資料庫中的 tb_admin_user 表，程式如下：

```
1.  // 實體，對應資料庫中的 tb_admin_user 表
2.  @Entity
3.  @Table(name = "tb_admin_user")
4.  class AdminUser {
5.      // 使用者 id
6.      @Id
7.      @GeneratedValue(strategy = GenerationType.IDENTITY)
```

```
8.        var adminUserId: Int? = null
9.        // 登入使用者名稱
10.       var loginUserName: String = ""
11.           set(loginUserName) {
12.               field = loginUserName?.trim { it <= ' ' }
13.           }
14.       // 登入密碼
15.       var loginPassword: String = ""
16.           set(loginPassword) {
17.               field = loginPassword?.trim { it <= ' ' }
18.           }
19.       // 暱稱
20.       var nickName: String = ""
21.           set(nickName) {
22.               field = nickName?.trim { it <= ' ' }
23.           }
24.       // 是否鎖定
25.       var locked: Byte = 0
26. }
```

Blog.kt 定義了部落格（文章）實體，包含 blogId、blogTitle、blogSubUrl、blogCoverImage、blogCategoryId、blogCategoryName、blogTags、blogStatus、blogViews、enableComment、isDeleted、createTime、updateTime、blogContent 這幾個屬性，對應資料庫中的 tb_blog 表。程式如下：

```
1.  // blog 實體，對應資料庫中的 tb_blog 表
2.  @Entity
3.  @Table(name = "tb_blog")
4.  class Blog {
5.      // 部落格 id
6.      @Id
7.      @GeneratedValue(strategy = GenerationType.IDENTITY)
8.      var blogId: Long = 0
```

```
9.        // 部落格標題
10.       var blogTitle: String = ""
11.           set(blogTitle) {
12.               field = blogTitle.trim { it <= ' ' }
13.           }
14.       // 部落格 url
15.       var blogSubUrl: String = ""
16.           set(blogSubUrl) {
17.               field = blogSubUrl.trim { it <= ' ' }
18.           }
19.       // 部落格封面圖片
20.       var blogCoverImage: String = ""
21.           set(blogCoverImage) {
22.               field = blogCoverImage.trim { it <= ' ' }
23.           }
24.       // 部落格分類 id
25.       var blogCategoryId: Int = 0
26.       // 部落格分類名稱
27.       var blogCategoryName: String = ""
28.           set(blogCategoryName) {
29.               field = blogCategoryName.trim { it <= ' ' }
30.           }
31.       // 部落格標籤
32.       var blogTags: String = ""
33.           set(blogTags) {
34.               field = blogTags.trim { it <= ' ' }
35.           }
36.       // 部落格審核狀態
37.       var blogStatus: Byte = 0
38.       // 部落格點擊量
39.       var blogViews: Long = 0
40.       // 部落格是否允許評論
41.       var enableComment: Byte = 0
42.       // 部落格是否被刪除
43.       var isDeleted: Byte = 0
```

```
44.      // 部落格建立時間
45.      @JsonFormat(pattern = "yyyy-MM-dd HH:mm:ss", timezone = "GMT+8")
46.      var createTime: Date = Date()
47.      // 部落格更新時間
48.      var updateTime: Date = Date()
49.      // 部落格內容
50.      var blogContent: String = ""
51.          set(blogContent) {
52.              field = blogContent.trim { it <= ' ' }
53.          }
54.  }
```

BlogCategory.kt 定義了部落格分類實體，包含 categoryId、categoryName、categoryIcon、categoryRank、isDeleted、createTime 屬性，對應資料庫中的 tb_blog_category 表。程式如下：

```
1.   // 部落格分類實體，對應資料庫中的 tb_blog_category 表
2.   @Entity
3.   @Table(name = "tb_blog_category")
4.   class BlogCategory {
5.       // 部落格分類 id
6.       @Id
7.       @GeneratedValue(strategy = GenerationType.IDENTITY)
8.       var categoryId: Int = 0
9.       // 部落格分類名稱
10.      var categoryName: String = ""
11.          set(categoryName) {
12.              field = categoryName.trim { it <= ' ' }
13.          }
14.      // 部落格分類圖示
15.      var categoryIcon: String = ""
16.          set(categoryIcon) {
17.              field = categoryIcon.trim { it <= ' ' }
18.          }
19.      // 部落格分類排名
```

```
20.      var categoryRank: Int = 0
21.      // 部落格分類是否被刪除
22.      var isDeleted: Byte = 0
23.      @JsonFormat(pattern = "yyyy-MM-dd HH:mm:ss", timezone = "GMT+8")
24.      // 建立時間
25.      var createTime: Date = Date()
26. }
```

BlogComment.kt 定義了部落格評論實體，包含 commentId、blogId、commentator、email、websiteUrl、commentBody、commentCreateTime、commentatorIp、replyBody、replyCreateTime、commentStatus、isDeleted 屬性，對應資料庫中的 tb_blog_comment 表。程式如下：

```
1.   // 部落格評論實體，對應資料庫表中的 tb_blog_comment 表
2.   @Entity
3.   @Table(name = "tb_blog_comment")
4.   class BlogComment {
5.       // 評論 id
6.       @Id
7.       @GeneratedValue(strategy = GenerationType.IDENTITY)
8.       var commentId: Long? = null
9.        // 部落格 id
10.      var blogId: Long = 0
11.       // 評論人
12.      var commentator: String = ""
13.          set(commentator) {
14.              field = commentator.trim { it <= ' ' }
15.          }
16.       // 電子郵件
17.      var email: String = ""
18.          set(email) {
19.              field = email.trim { it <= ' ' }
20.          }
21.       // 網址
22.      var websiteUrl: String = ""
```

```
23.         set(websiteUrl) {
24.             field = websiteUrl.trim { it <= ' ' }
25.         }
26.     // 評論內容
27.     var commentBody: String = ""
28.         set(commentBody) {
29.             field = commentBody.trim { it <= ' ' }
30.         }
31.     // 評論建立時間
32.     @JsonFormat(pattern = "yyyy-MM-dd HH:mm:ss", timezone = "GMT+8")
33.     var commentCreateTime: Date = Date()
34.     // 評論者 IP 位址
35.     var commentatorIp: String = ""
36.         set(commentatorIp) {
37.             field = commentatorIp.trim { it <= ' ' }
38.         }
39.     // 回覆內容
40.     var replyBody: String = ""
41.         set(replyBody) {
42.             field = replyBody.trim { it <= ' ' }
43.         }
44.     // 回覆建立時間
45.     @JsonFormat(pattern = "yyyy-MM-dd HH:mm:ss", timezone = "GMT+8")
46.     var replyCreateTime: Date = Date()
47.     // 評論狀態
48.     var commentStatus: Byte = 0
49.     // 是否刪除
50.     var isDeleted: Byte = 0
51. }
```

BlogConfig.kt 定義了部落格設定實體，包含 configName、configValue、createTime、updateTime 屬性，對應資料庫中的 tb_config 表。程式如下：

```
1. // 部落格設定實體，對應資料庫中的 tb_config 表
2. @Entity
```

```kotlin
3.    @Table(name = "tb_config")
4.    class BlogConfig {
5.        // 設定名稱
6.        @Id
7.        var configName: String = ""
8.            set(configName) {
9.                field = configName.trim { it <= ' ' }
10.           }
11.       // 設定值
12.       var configValue: String = ""
13.           set(configValue) {
14.               field = configValue.trim { it <= ' ' }
15.           }
16.       // 設定建立時間
17.       @JsonFormat(pattern = "yyyy-MM-dd HH:mm:ss", timezone = "GMT+8")
18.       var createTime: Date = Date()
19.       // 設定更新時間
20.       @JsonFormat(pattern = "yyyy-MM-dd HH:mm:ss", timezone = "GMT+8")
21.       var updateTime: Date = Date()
22.   }
```

BlogLink.kt 定義了部落格連結（友情連結）實體，包含 linkId、linkType、linkName、linkUrl、linkDescription、linkRank、isDeleted、createTime 屬性，對應資料庫中的 tb_link 表。程式如下：

```kotlin
1.    // 友情連結實體，對應資料庫中的 tb_link 表
2.    @Entity
3.    @Table(name = "tb_link")
4.    class BlogLink {
5.        // 友情連結 id
6.        @Id
7.        @GeneratedValue(strategy = GenerationType.IDENTITY)
8.        var linkId: Int? = null
9.        // 友情連結類型
10.       var linkType: Byte = 0
```

```
11.    // 友情連結名稱
12.    var linkName: String = ""
13.        set(linkName) {
14.            field = linkName.trim { it <= ' ' }
15.        }
16.    // 友情連結位址
17.    var linkUrl: String = ""
18.        set(linkUrl) {
19.            field = linkUrl.trim { it <= ' ' }
20.        }
21.    // 友情連結描述
22.    var linkDescription: String = ""
23.        set(linkDescription) {
24.            field = linkDescription.trim { it <= ' ' }
25.        }
26.    // 友情連結排名
27.    var linkRank: Int = 0
28.    // 友情連結是否刪除
29.    var isDeleted: Byte = 0
30.    // 友情連結建立時間
31.    @JsonFormat(pattern = "yyyy-MM-dd HH:mm:ss", timezone = "GMT+8")
32.    var createTime: Date = Date()
33. }
```

BlogTag.kt 定義了部落格標籤實體，包含 tagId、tagName、isDeleted、createTime 屬性，對應資料庫中的 tb_blog_tag 表。程式如下：

```
1.  // 標籤實體，對應資料庫中的 tb_blog_tag 表
2.  @Entity
3.  @Table(name = "tb_blog_tag")
4.  class BlogTag {
5.      // 標籤 id
6.      @Id
7.      @GeneratedValue(strategy = GenerationType.IDENTITY)
8.      var tagId: Int = 0
```

```
9.      // 標籤名稱
10.     var tagName: String = ""
11.        set(tagName) {
12.            field = tagName.trim { it <= ' ' }
13.        }
14.     // 標籤是否刪除
15.     var isDeleted: Byte = 0
16.     // 標籤建立時間
17.     @JsonFormat(pattern = "yyyy-MM-dd HH:mm:ss", timezone = "GMT+8")
18.     var createTime: Date = Date()
19. }
```

BlogTagRelation.kt 定義了部落格標籤關係實體，包含 relationId、blogId、tagId、createTime 屬性，對應資料庫中的 tb_blog_tag_relation 表。程式如下：

```
1.  // 部落格標籤關係實體，對應資料庫 tb_blog_tag_relation 表
2.  @Entity
3.  @Table(name = "tb_blog_tag_relation")
4.  class BlogTagRelation {
5.      // 部落格標籤關係 id
6.      @Id
7.      @GeneratedValue(strategy = GenerationType.IDENTITY)
8.      var relationId: Long? = null
9.      // 部落格 id
10.     var blogId: Long = 0
11.     // 標籤 id
12.     var tagId: Int = 0
13.     // 建立時間
14.     @JsonFormat(pattern = "yyyy-MM-dd HH:mm:ss", timezone = "GMT+8")
15.     var createTime: Date = Date()
16. }
```

10.4 資料庫設計

10.3 節中的實體對應不同的資料庫表：tb_admin_user、tb_blog、tb_blog_category、tb_blog_comment、tb_config、tb_link、tb_blog_tag、tb_blog_tag_relation。資料庫表的關係如圖 10.2 所示。

圖 10.2 資料庫圖表

10.5 Repository 層的設計

部落格系統定義的 Repository/DAO 如表 10.3 所示。

表 10.3 部落格定義的 Repository/DAO 列表

序號	Repository/DAO	說明	方法
1	AdminUserRepository	管理使用者 JPA 介面	提供增刪改查方法
2	BlogCategoryRepository	分類 JPA 介面	提供增刪改查方法
3	BlogCommentRepository	評論 JPA 介面	提供增刪改查方法
4	BlogConfigRepository	設定 JPA 介面	提供增刪改查方法
5	BlogLinkRepository	連結 JPA 介面	提供增刪改查方法
6	BlogRepository	部落格 JPA 介面	提供增刪改查方法
7	BlogTagRelationRepository	部落格標籤關係 JPA 介面	提供增刪改查方法
8	BlogTagRepository	標籤 JPA 介面	提供增刪改查方法
9	AdminUserDAO	使用者 DAO 類別	新增使用者、登入、根據 userId 查詢、更新記錄
10	BlogCategoryDAO	分類 DAO 類別	邏輯刪除、新增分類、查詢分類、更新欄位、分頁查詢、批次查詢、批次刪除、查詢總數
11	BlogCommentDAO	評論 DAO 類別	邏輯刪除、新增評論、查詢評論、選擇性更新、分頁尋找、查詢總數、批次刪除
12	BlogConfigDAO	設定 DAO 類別	查詢所有設定、根據設定名稱查詢、選擇性更新

序號	Repository/DAO	說明	方法
13	BlogDAO	部落格 DAO 類別	刪除、儲存部落格、根據部落格 id 查詢部落格、選擇性更新部落格、分頁查詢部落格、根據類型查詢部落格、查詢總數、批次刪除、根據標籤分頁查詢、更新分類
14	BlogLinkDAO	連結 DAO 類別	刪除友情連結、新增友情連結、尋找友情連結、選擇性更新友情連結、分頁尋找友情連結、查詢總數、批次刪除
15	BlogTagDAO	標籤 DAO 類別	刪除標籤、新增標籤、查詢標籤、選擇性更新標籤、分頁查詢標籤、查詢總數、批次刪除
16	BlogTagRelationDAO	部落格標籤關係 DAO 類別	刪除、儲存、查詢、選擇性更新、批次插入、批次刪除

AdminUserRepository.kt、BlogCategoryRepository.kt、BlogComment
Repository.kt、BlogConfigRepository.kt、BlogLinkRepository.kt、
BlogRepository.kt、BlogTagRelationRepository.kt、BlogTagRepository.kt 都
繼承了 JpaRepository，JPA 提供了增刪改查方法。但是 JPA 對原生 SQL 支
援不太好，QueryDSL 提供了便利。

```
1.  // 操作 AdminUser 對應的資料庫表的 JpaRepository 介面
2.  interface AdminUserRepository : JpaRepository<AdminUser, Long>,
    QuerydslPredicateExecutor<AdminUser>
3.  // 操作 BlogCategory 對應的資料庫表的 JpaRepository 介面
4.  interface BlogCategoryRepository : JpaRepository<BlogCategory, Long>,
    QuerydslPredicateExecutor<BlogCategory>
5.  // 操作 BlogComment 對應的資料庫表的 JpaRepository 介面
6.  interface BlogCommentRepository : JpaRepository<BlogComment, Long>,
    QuerydslPredicateExecutor<BlogComment>
```

```
7.  // 操作 BlogConfig 對應的資料庫表的 JpaRepository 介面
8.  interface BlogConfigRepository : JpaRepository<BlogConfig, String>,
    QuerydslPredicateExecutor<BlogConfig>
9.  // 操作 BlogLink 對應的資料庫表的 JpaRepository 介面
10. interface BlogLinkRepository : JpaRepository<BlogLink, Long>,
    QuerydslPredicateExecutor<BlogLink>
11. // 操作 Blog 對應的資料庫表的 JpaRepository 介面
12. interface BlogRepository : JpaRepository<Blog, Long>,
    QuerydslPredicateExecutor<Blog>
13. // 操作 BlogTagRelation 對應的資料庫表的 JpaRepository 介面
14. interface BlogTagRelationRepository : JpaRepository<BlogTagRelation,
    Long>, QuerydslPredicateExecutor<BlogTagRelation>
15. // 操作 BlogTag 對應的資料庫表的 JpaRepository 介面
16. interface BlogTagRepository : JpaRepository<BlogTag, Long>,
    QuerydslPredicateExecutor<BlogTag>
```

AdminUserDAO.kt 基於 QueryDSL 實現了新增使用者；根據使用者名稱、密碼驗證該使用者是否存在，模擬登入行為；根據使用者 id 查詢使用者；對欄位進行選擇性更新，當欄位不為 null 才會更新相關記錄的欄位；根據使用者 id，更新指定使用者的欄位。程式如下：

```
1.  @Component
2.  class AdminUserDAO {
3.      @Autowired
4.      lateinit var adminUserRepository: AdminUserRepository
5.      @Autowired
6.      lateinit var queryFactory: JPAQueryFactory
7.      @Transactional
8.      // 儲存一筆 AdminUser 資料
9.      fun insert(record: AdminUser): Int {
10.         adminUserRepository.save(record)
11.         return 0
12.     }
13.     // 儲存一筆 AdminUser 資料
14.     @Transactional
```

```
15.    fun insertSelective(record: AdminUser): Int {
16.        adminUserRepository.save(record)
17.        return 0
18.    }
19.    // 根據 userName、password 查詢是否存在一筆 AdminUser
20.    fun login(userName: String, password: String): AdminUser? {
21.        val qAdminUser = QAdminUser.adminUser
22.        return queryFactory.selectFrom(qAdminUser)
23.                .where(qAdminUser.loginUserName.eq(userName).and
    (qAdminUser.loginPassword.eq(password)))
24.                .fetchOne()
25.    }
26.    // 根據 adminUserId 尋找 AdminUser
27.    fun selectByPrimaryKey(adminUserId: Int): AdminUser? {
28.        val qAdminUser = QAdminUser.adminUser
29.        return queryFactory.selectFrom(qAdminUser)
30.                .where(qAdminUser.adminUserId.eq(adminUserId))
31.                .fetchOne()
32.    }
33.    // 將 record 中不為 null 的欄位更新到資料庫
34.    @Transactional
35.    fun updateByPrimaryKeySelective(record: AdminUser): Int {
36.        val qAdminUser = QAdminUser.adminUser
37.        val cols = arrayListOf<Path<*>>()
38.        val values = arrayListOf<Any?>()
39.        if(record.loginUserName != null) {
40.            cols.add(qAdminUser.loginUserName)
41.            values.add(record.loginUserName)
42.        }
43.        if(record.loginPassword != null) {
44.            cols.add(qAdminUser.loginPassword)
45.            values.add(record.loginPassword)
46.        }
47.        if(record.nickName != null) {
48.            cols.add(qAdminUser.nickName)
```

```
49.              values.add(record.nickName)
50.         }
51.         if(record.locked != null) {
52.              cols.add(qAdminUser.locked)
53.              values.add(record.locked)
54.         }
55.         return queryFactory.update(qAdminUser)
56.                  .set(cols, values)
57.                  .where(qAdminUser.adminUserId.eq(record.adminUserId))
58.                  .execute()
59.                  .toInt()
60.      }
61.      // 更新 AdminUser 資料
62.      @Transactional
63.      fun updateByPrimaryKey(record: AdminUser): Int {
64.          val qAdminUser = QAdminUser.adminUser
65.          return queryFactory.update(qAdminUser)
66.                  .set(qAdminUser.loginUserName, record.loginUserName)
67.                  .set(qAdminUser.loginPassword, record.loginPassword)
68.                  .set(qAdminUser.nickName, record.nickName)
69.                  .set(qAdminUser.locked, record.locked)
70.                  .where(qAdminUser.adminUserId.eq(record.adminUserId))
71.                  .execute()
72.                  .toInt()
73.      }
74. }
```

BlogCategoryDAO.kt 基於 QueryDSL 實現了根據分類 id 刪除該筆記錄，把該記錄的 isDeleted 更新為 1，表示刪除；插入一筆 BlogCategory；根據 categoryId 查詢 BlogCategory；根據 categoryName 查詢 BlogCategory；BlogCategory 屬性不為 null，更新資料庫對應欄位；根據 categoryId 更新該分類相關欄位；按照 categoryRank、createTime 降冪分頁查詢有效的部落格分類；根據 categoryId 批次查詢分類；查詢分類總數；將批次更新記錄的 isDeleted 設為 1。程式如下：

```
1.  @Component
2.  class BlogCategoryDAO {
3.      @Autowired
4.      lateinit var queryFactory: JPAQueryFactory
5.      @Autowired
6.      lateinit var blogCategoryRepository: BlogCategoryRepository
7.      // 根據 categoryId 刪除
8.      @Transactional
9.      fun deleteByPrimaryKey(categoryId: Int): Int {
10.         // 省略部分程式……
11.         return queryFactory.update(qBlogCategory)
12.                 .set(qBlogCategory.isDeleted, 1)
13.                 .where(predicate)
14.                 .execute()
15.                 .toInt()
16.     }
17.     // 儲存一筆 BlogCategory 資料
18.     @Transactional
19.     fun insert(record: BlogCategory): Int {
20.         blogCategoryRepository.save(record)
21.         return 0
22.     }
23.     // 儲存一筆 BlogCategory 資料
24.     @Transactional
25.     fun insertSelective(record: BlogCategory): Int {
26.         blogCategoryRepository.save(record)
27.         return 0
28.     }
29.     // 根據 categoryId 尋找 BlogCategory
30.     fun selectByPrimaryKey(categoryId: Int?): BlogCategory? {
31.         // 省略部分程式……
32.         val predicate = qBlogCategory.categoryId.eq(categoryId)
33.                 .and(qBlogCategory.isDeleted.eq(0))
34.     }
35.     // 根據 categoryName 尋找 BlogCategory
```

```kotlin
36.    fun selectByCategoryName(categoryName: String): BlogCategory? {
37.        // 省略部分程式……
38.        val predicate = qBlogCategory.categoryName.eq(categoryName)
39.                .and(qBlogCategory.isDeleted.eq(0))
40.    }
41.    // 將 record 不為 null 的欄位更新到資料庫
42.    @Transactional
43.    fun updateByPrimaryKeySelective(record: BlogCategory): Int {
44.        // 省略部分程式……
45.        return queryFactory.update(qBlogCategory)
46.                .set(cols, values)
47.                .where(qBlogCategory.categoryId.eq(record.categoryId))
48.                .execute()
49.                .toInt()
50.    }
51.    // 更新 BlogCategory
52.    @Transactional
53.    fun updateByPrimaryKey(record: BlogCategory): Int {
54.        val qBlogCategory = QBlogCategory.blogCategory
55.        return queryFactory.update(qBlogCategory)
56.                .set(qBlogCategory.categoryName, record.categoryName)
57.                .set(qBlogCategory.categoryIcon, record.categoryIcon)
58.                .set(qBlogCategory.categoryRank, record.categoryRank)
59.                .set(qBlogCategory.isDeleted, record.isDeleted)
60.                .set(qBlogCategory.createTime, record.createTime)
61.                .where(qBlogCategory.categoryId.eq(record.categoryId))
62.                .execute()
63.                .toInt()
64.    }
65.    // 分頁尋找 BlogCategory
66.    fun findCategoryList(pageUtil: PageQueryUtil?): List<BlogCategory> {
67.        // 省略部分程式……
68.    }
69.    // 根據 categoryIds 尋找 BlogCategory
70.    fun selectByCategoryIds(categoryIds: List<Int?>): List<BlogCategory> {
```

```
71.        // 省略部分程式……
72.        val predicate = qBlogCategory.categoryId.`in`(categoryIds)
73.                .and(qBlogCategory.isDeleted.eq(0))
74.    }
75.    // 查詢總數
76.    fun getTotalCategories(pageUtil: PageQueryUtil?): Int {
77.        // 省略部分程式……
78.        return queryFactory.selectFrom(qBlogCategory)
79.                .where(qBlogCategory.isDeleted.eq(0))
80.                .fetchCount()
81.                .toInt()
82.    }
83.    // 批次刪除 BlogCategory
84.    @Transactional
85.    fun deleteBatch(ids: List<Int>): Int {
86.        // 省略部分程式……
87.        val predicate = qBlogCategory.categoryId.`in`(ids)
88.        return queryFactory.update(qBlogCategory)
89.                .set(qBlogCategory.isDeleted, 1)
90.                .where(predicate)
91.                .execute()
92.                .toInt()
93.    }
94. }
```

BlogCommentDAO.kt 基於 QueryDSL 實現了根據 commentId 物理刪除評論；插入一筆評論；根據 commentId 查詢該評論；當欄位不為 null 時更新資料庫對應欄位；根據 commentId 更新對應的記錄；按照 commentId 降冪分頁尋找評論；根據 blogId、commentStatus 查詢評論總數；將一組評論的 commentStatus 更新為 1；批次刪除評論。程式如下：

```
1.  @Component
2.  class BlogCommentDAO {
3.      @Autowired
4.      lateinit var queryFactory: JPAQueryFactory
```

```kotlin
5.      @Autowired
6.      lateinit var blogCommentRepository: BlogCommentRepository
7.      // 根據 commentId 刪除
8.      @Transactional
9.      fun deleteByPrimaryKey(commentId: Long): Int {
10.         // 省略部分程式……
11.     }
12.     // 儲存一筆 BlogComment
13.     @Transactional
14.     fun insert(record: BlogComment): Int {
15.         blogCommentRepository.save(record)
16.         return 0
17.     }
18.     // 儲存一筆 BlogComment
19.     @Transactional
20.     fun insertSelective(record: BlogComment): Int {
21.         blogCommentRepository.save(record)
22.         return 0
23.     }
24.     // 根據 commentId 尋找 BlogComment
25.     fun selectByPrimaryKey(commentId: Long): BlogComment? {
26.         val qBlogComment = QBlogComment.blogComment
27.         return queryFactory.selectFrom(qBlogComment)
28.                 .where(qBlogComment.commentId.eq(commentId).and
    (qBlogComment.isDeleted.eq(0)))
29.                 .fetchOne()
30.     }
31.     // 將 record 中不為 null 的欄位更新到資料庫
32.     @Transactional
33.     fun updateByPrimaryKeySelective(record: BlogComment): Int {
34.         // 省略部分程式……
35.         return queryFactory.update(qBlogComment)
36.                 .set(cols, values)
37.                 .where(qBlogComment.commentId.eq(record.commentId))
38.                 .execute()
```

```
39.                    .toInt()
40.        }
41.        // 更新 BlogComment
42.        @Transactional
43.        fun updateByPrimaryKey(record: BlogComment): Int {
44.            // 省略部分程式……
45.        }
46.        // 分頁尋找 BlogComment
47.        fun findBlogCommentList(map: Map<*, *>): List<BlogComment> {
48.            // 省略部分程式……
49.            return if(start != null && limit != null) {
50.                queryFactory.selectFrom(qBlogComment)
51.                        .where(predicate)
52.                        .where(predicate1)
53.                        .where(qBlogComment.isDeleted.eq(0))
54.                        .orderBy(OrderSpecifier(Order.DESC, qBlogComment.
    commentId))
55.                        .offset(start.toLong())
56.                        .limit(limit.toLong())
57.                        .fetchResults()
58.                        .results
59.            }else {
60.                listOf()
61.            }
62.        }
63.        // 根據 blogId、CommentStatus 查詢總數
64.        fun getTotalBlogComments(map: Map<*, *>?): Int {
65.            // 省略部分程式……
66.          return queryFactory.selectFrom(qBlogComment)
67.                        .where(predicate)
68.                        .where(predicate1)
69.                        .where(qBlogComment.isDeleted.eq(0))
70.                        .fetchCount()
71.                        .toInt()
72.        }
```

```
73.      // 根據 ids，將 commentStatus 設定為 1
74.      @Transactional
75.      fun checkDone(ids: List<Long>): Int {
76.          val qBlogComment = QBlogComment.blogComment
77.          return queryFactory.update(qBlogComment)
78.                  .set(qBlogComment.commentStatus, 1)
79.                  .where(qBlogComment.commentId.`in`(ids).and (qBlogComment.
    commentStatus.eq(0)))
80.                  .execute()
81.                  .toInt()
82.      }
83.      // 根據 ids，批次刪除
84.      @Transactional
85.      fun deleteBatch(ids: List<Long>): Int {
86.          val qBlogComment = QBlogComment.blogComment
87.          return queryFactory.update(qBlogComment)
88.                  .set(qBlogComment.isDeleted, 1)
89.                  .where(qBlogComment.commentId.`in`(ids))
90.                  .execute()
91.                  .toInt()
92.      }
93. }
```

BlogConfigDAO.kt 基於 QueryDSL 實現了查詢所有部落格設定；根據設定名稱查詢對應的設定；選擇性更新不為 null 的設定屬性等。程式如下：

```
1.  @Component
2.  class BlogConfigDAO {
3.      @Autowired
4.      lateinit var queryFactory: JPAQueryFactory
5.      // 尋找所有資料
6.      fun selectAll(): List<BlogConfig> {
7.          val qBlogConfig = QBlogConfig.blogConfig
8.          return queryFactory.selectFrom(qBlogConfig)
9.                  .fetchResults()
```

```
10.                    .results
11.        }
12.        // 根據 configName 尋找資料
13.        fun selectByPrimaryKey(configName: String): BlogConfig {
14.            val qBlogConfig = QBlogConfig.blogConfig
15.            val predicate = qBlogConfig.configName.eq(configName)
16.            return queryFactory.selectFrom(qBlogConfig)
17.                    .where(predicate)
18.                    .fetchFirst()
19.        }
20.        // 將 record 不為 null 的欄位更新到資料庫
21.        @Transactional
22.        fun updateByPrimaryKeySelective(record: BlogConfig): Int {
23.            // 省略部分程式……
24.            return queryFactory.update(qBlogConfig)
25.                    .set(cols, values)
26.                    .where(qBlogConfig.configName.eq(record.configName))
27.                    .execute()
28.                    .toInt()
29.        }
30. }
```

BlogDAO.kt 基於 QueryDSL 實現了刪除指定 blogId 的部落格；儲存一筆部落格；根據 blogId 查詢部落格；選擇性更新不為 null 的部落格屬性；根據 blogId 更新指定部落格的屬性；根據關鍵字、狀態、分類按照 blogId 降冪分頁尋找部落格；根據 blogId 或 blogViews 降冪查詢部落格；根據關鍵字、狀態、分類查詢部落格總數；批次刪除部落格；按照 blogId 降冪分頁尋找指定標籤的部落格；查詢指定標籤的部落格總數；根據 url 查詢部落格；更新部落格的標籤和分類。程式如下：

```
1.  @Component
2.  class BlogDAO {
3.      @Autowired
4.      lateinit var queryFactory: JPAQueryFactory
```

```
5.      @Autowired
6.      lateinit var blogRepository: BlogRepository
7.      // 根據 blogId 刪除資料
8.      @Transactional
9.      fun deleteByPrimaryKey(blogId: Long): Int {
10.         val qBlog = QBlog.blog
11.         return queryFactory.update(qBlog)
12.             .set(qBlog.isDeleted, 1)
13.             .where(qBlog.isDeleted.eq(0).and(qBlog.blogId.eq(blogId)))
14.             .execute()
15.             .toInt()
16.     }
17.     // 儲存一筆 Blog
18.     @Transactional
19.     fun insert(record: Blog): Int {
20.         blogRepository.save(record)
21.         return 0
22.     }
23.     // 儲存一筆 Blog
24.     @Transactional
25.     fun insertSelective(record: Blog): Int {
26.         blogRepository.save(record)
27.         return 0
28.     }
29.     // 根據 blogId 尋找 Blog
30.     fun selectByPrimaryKey(blogId: Long?): Blog {
31.         val qBlog = QBlog.blog
32.         return queryFactory.selectFrom(qBlog)
33.             .where(qBlog.blogId.eq(blogId))
34.             .fetchFirst()
35.     }
36.     // 將 record 不為 null 的欄位更新到資料庫
37.     @Transactional
38.     fun updateByPrimaryKeySelective(record: Blog): Int {
39.         // 省略部分程式……
```

```
40.        return queryFactory.update(qBlog)
41.                .set(cols, values)
42.                .where(qBlog.blogId.eq(record.blogId))
43.                .execute()
44.                .toInt()
45.    }
46.    // 更新 Blog
47.    @Transactional
48.    fun updateByPrimaryKeyWithBLOBs(record: Blog): Int {
49.        // 省略部分程式……
50.    }
51.    // 更新 Blog
52.    @Transactional
53.    fun updateByPrimaryKey(record: Blog): Int {
54.        // 省略部分程式……
55.    }
56.    // 分頁尋找 Blog
57.    fun findBlogList(pageUtil: PageQueryUtil): List<Blog> {
58.        // 省略部分程式……
59.        return queryFactory.selectFrom(qBlog)
60.                .where(predicate)
61.                .where(predicate1)
62.                .where(predicate2)
63.                .where(qBlog.isDeleted.eq(0))
64.                .orderBy(OrderSpecifier(Order.DESC, qBlog.blogId))
65.                .offset(start.toLong())
66.                .limit(limit.toLong())
67.                .fetchResults()
68.                .results
69.    }
70.    // 根據 blogId 或 blogViews 降冪尋找 limit 筆 Blog
71.    fun findBlogListByType(type: Int, limit: Int): List<Blog> {
72.        // 省略部分程式……
73.        return queryFactory.selectFrom(qBlog)
74.                .where(qBlog.isDeleted.eq(0).and(qBlog.blogStatus.eq(1)))
```

```
75.                    .orderBy(order)
76.                    .limit(limit.toLong())
77.                    .fetchResults().results
78.          }
79.     // 查詢滿足條件的 blog 總數
80.     fun getTotalBlogs(pageUtil: PageQueryUtil?): Int {
81.          // 省略部分程式……
82.          return queryFactory.selectFrom(qBlog)
83.                    .where(predicate)
84.                    .where(predicate1)
85.                    .where(predicate2)
86.                    .where(qBlog.isDeleted.eq(0))
87.                    .fetchCount().toInt()
88.          }
89.     // 根據 ids 批次刪除 blog
90.     @Transactional
91.     fun deleteBatch(ids: List<Long>): Int {
92.          val qBlog = QBlog.blog
93.          return queryFactory.update(qBlog)
94.                    .set(qBlog.isDeleted, 1)
95.                    .where(qBlog.blogId.`in`(ids))
96.                    .execute()
97.                    .toInt()
98.          }
99.     // 根據 tagId，分頁尋找 Blog
100.     fun getBlogsPageByTagId(pageUtil: PageQueryUtil): List<Blog> {
101.          // 省略部分程式……
102.          return queryFactory.selectFrom(qBlog)
103.                    .where(qBlog.blogStatus.eq(1).and(qBlog.isDeleted.eq(0)))
104.                    .where(qBlog.blogId.`in`(blogIds))
105.                    .orderBy(OrderSpecifier(Order.DESC, qBlog.blogId))
106.                    .offset(start.toLong())
107.                    .limit(limit.toLong())
108.                    .fetchResults()
109.                    .results
```

```
110.        }
111.        // 根據 tagId 查詢總數
112.        fun getTotalBlogsByTagId(pageUtil: PageQueryUtil): Int {
113.            // 省略部分程式……
114.            return queryFactory.selectFrom(qBlog)
115.                    .where(qBlog.blogStatus.eq(1).and(qBlog.isDeleted.eq(0)))
116.                    .where(qBlog.blogId.`in`(blogIds))
117.                    .fetchCount().toInt()
118.        }
119.        // 根據 subUrl 尋找 Blog
120.        fun selectBySubUrl(subUrl: String): Blog? {
121.            val qBlog = QBlog.blog
122.            return queryFactory.selectFrom(qBlog)
123.                    .where(qBlog.blogSubUrl.eq(subUrl).and(qBlog.
     isDeleted.eq(0)))
124.                    .limit(1)
125.                    .fetchOne()
126.        }
127.        // 批次更新 categoryName 和 categoryId
128.        @Transactional
129.        fun updateBlogCategorys(categoryName: String, categoryId: Int,
     ids: List<Long>): Int {
130.            val qBlog = QBlog.blog
131.            return queryFactory.update(qBlog)
132.                    .set(qBlog.blogCategoryId, categoryId)
133.                    .set(qBlog.blogCategoryName, categoryName)
134.                    .where(qBlog.blogId.`in`(ids).and(qBlog.isDeleted.
     eq(0)))
135.                    .execute().toInt()
136.        }
137.    }
```

BlogLinkDAO.kt 基於 QueryDSL 實現了根據 linkId 刪除連結（友情連結）；
插入一條友情連結；根據 linkId 查詢友情連結；更新不為 null 的友情連結
屬性；根據友情連結 id 更新相關屬性；按照友情連結 id 降冪分頁尋找友

情連結;查詢友情連結總數;批次刪除友情連結等。程式如下:

```kotlin
1.  @Component
2.  class BlogLinkDAO {
3.      @Autowired
4.      lateinit var queryFactory: JPAQueryFactory
5.      @Autowired
6.      lateinit var blogLinkRepository: BlogLinkRepository
7.      // 根據 linkId 刪除 BlogLink
8.      @Transactional
9.      fun deleteByPrimaryKey(linkId: Int): Int {
10.         val qBlogLink = QBlogLink.blogLink
11.         return queryFactory.update(qBlogLink)
12.             .set(qBlogLink.isDeleted, 1)
13.             .where(qBlogLink.linkId.eq(linkId).and(qBlogLink.isDeleted.eq(0)))
14.             .execute()
15.             .toInt()
16.     }
17.     // 儲存一筆 BlogLink
18.     @Transactional
19.     fun insert(record: BlogLink): Int {
20.         blogLinkRepository.save(record)
21.         return 0
22.     }
23.     // 儲存一筆 BlogLink
24.     @Transactional
25.     fun insertSelective(record: BlogLink): Int {
26.         blogLinkRepository.save(record)
27.         return 0
28.     }
29.     // 根據 linkId 尋找 BlogLink
30.     fun selectByPrimaryKey(linkId: Int?): BlogLink {
31.         val qBloglink = QBlogLink.blogLink
32.         return queryFactory.selectFrom(qBloglink)
```

```
33.                    .where(qBloglink.linkId.eq(linkId).and(qBloglink.
    isDeleted.eq(0)))
34.                    .fetchFirst()
35.       }
36.       // 將 record 不為 null 的欄位更新到資料庫
37.       @Transactional
38.       fun updateByPrimaryKeySelective(record: BlogLink): Int {
39.           // 省略部分程式……
40.           return queryFactory.update(qBlogLink)
41.                    .set(cols, values)
42.                    .where(qBlogLink.linkId.eq(record.linkId))
43.                    .execute()
44.                    .toInt()
45.       }
46.       // 更新 BlogLink
47.       @Transactional
48.       fun updateByPrimaryKey(record: BlogLink): Int {
49.           // 省略部分程式……
50.       }
51.       // 分頁尋找 BlogLink
52.       fun findLinkList(pageUtil: PageQueryUtil?): List<BlogLink> {
53.           // 省略部分程式……
54.       return return queryFactory.selectFrom(qBlogLink)
55.                    .where(qBlogLink.isDeleted.eq(0))
56.                    .orderBy(OrderSpecifier(Order.DESC, qBlogLink.linkId))
57.                    .fetchResults()
58.                    .results
59.       }
60.       // 查詢總數
61.       fun getTotalLinks(pageUtil: PageQueryUtil?): Int {
62.           // 省略部分程式……
63.       }
64.       // 根據 ids 批次刪除
65.       @Transactional
66.       fun deleteBatch(ids: List<Int>): Int {
```

```
67.        // 省略部分程式……
68.    }
69. }
```

BlogTagDAO.kt 基於 QueryDSL 實現了根據標籤 id 刪除標籤；插入一筆標籤；根據 tagId 查詢標籤；根據 tagName 查詢標籤；選擇性更新不為 null 的標籤屬性；根據 tagId 更新標籤名稱、建立時間；按照 tagId 降冪分頁尋找友情連結；查詢不同標籤對應的部落格的數量；查詢標籤總數；批次刪除標籤；批次儲存標籤等。程式如下：

```
1.  @Component
2.  class BlogTagDAO {
3.      @Autowired
4.      lateinit var queryFactory: JPAQueryFactory
5.      @Autowired
6.      lateinit var blogTagRepository: BlogTagRepository
7.      // 根據 tagId 刪除 BlogTag
8.      @Transactional
9.      fun deleteByPrimaryKey(tagId: Int): Int {
10.         val qBlogTag = QBlogTag.blogTag
11.         return queryFactory.update(qBlogTag)
12.                 .set(qBlogTag.isDeleted, 1)
13.                 .where(qBlogTag.tagId.eq(tagId))
14.                 .execute().toInt()
15.     }
16.     // 儲存一筆 BlogTag
17.     @Transactional
18.     fun insert(record: BlogTag): Int {
19.         blogTagRepository.save(record)
20.         return 0
21.     }
22.     // 儲存一筆 BlogTag
23.     @Transactional
24.     fun insertSelective(record: BlogTag): Int {
25.         blogTagRepository.save(record)
```

```
26.        return 0
27.    }
28.    // 根據 tagId 查詢 BlogTag
29.    fun selectByPrimaryKey(tagId: Int): BlogTag {
30.        val qBlogTag = QBlogTag.blogTag
31.        return queryFactory.selectFrom(qBlogTag)
32.            .where(qBlogTag.tagId.eq(tagId).and(qBlogTag.isDeleted.
    eq(0)))
33.            .fetchFirst()
34.    }
35.    // 根據 tagName 查詢 BlogTag
36.    fun selectByTagName(tagName: String): BlogTag {
37.        val qBlogTag = QBlogTag.blogTag
38.        return queryFactory.selectFrom(qBlogTag)
39.            .where(qBlogTag.tagName.eq(tagName).and(qBlogTag.
    isDeleted.eq(0)))
40.            .fetchFirst()
41.    }
42.    // 將 record 不為 null 的欄位更新到資料庫
43.    @Transactional
44.    fun updateByPrimaryKeySelective(record: BlogTag): Int {
45.        // 省略部分程式……
46.        return queryFactory.update(qBlogTag)
47.            .set(cols, values)
48.            .where(qBlogTag.tagId.eq(record.tagId))
49.            .execute()
50.            .toInt()
51.    }
52.    // 更新 BlogTag
53.    @Transactional
54.    fun updateByPrimaryKey(record: BlogTag): Int {
55.        // 省略部分程式……
56.    }
57.    // 分頁尋找 BlogTag
58.    fun findTagList(pageUtil: PageQueryUtil): List<BlogTag> {
```

```
59.        // 省略部分程式……
60.    }
61.    // 查詢每個 tagId、tagName 對應的 Blog 數量
62.    fun getTagCount(): List<BlogTagCount> {
63.        // 省略部分程式……
64.        return result.map {
65.            val blogTagCount = BlogTagCount()
66.            blogTagCount.tagId = it.get(0, Int::class.java)!!
67.            blogTagCount.tagName = it.get(1, String::class.java)!!
68.            blogTagCount.tagCount = it.get(2, Long::class.java)?.toInt()!!
69.            blogTagCount
70.        }.toList()
71.    }
72.    // 查詢 BlogTag 總數
73.    fun getTotalTags(pageUtil: PageQueryUtil?): Int {
74.        val qBlogTag = QBlogTag.blogTag
75.        return queryFactory.selectFrom(qBlogTag)
76.                .fetchCount().toInt()
77.    }
78.    // 根據 ids 批次刪除 BlogTag
79.    @Transactional
80.    fun deleteBatch(ids: List<Int>): Int {
81.        val qBlogTag = QBlogTag.blogTag
82.        return queryFactory.update(qBlogTag)
83.                .set(qBlogTag.isDeleted, 1)
84.                .where(qBlogTag.tagId.`in`(ids))
85.                .execute().toInt()
86.    }
87.    // 批次儲存 BlogTag
88.    @Transactional
89.    fun batchInsertBlogTag(tagList: List<BlogTag>): Int {
90.        blogTagRepository.saveAll(tagList)
91.        return 0
92.    }
93. }
```

BlogTagRelationDAO.kt 基於 QueryDSL 實現了根據 relationId 刪除記錄；
插入記錄；根據 relationId 尋找記錄；根據 blogId、tagId 尋找關係記錄；
根據標籤清單查詢關係記錄；選擇性更新不為 null 的記錄的屬性；更新記
錄的屬性；批次儲存關係記錄；批次刪除關係記錄等。程式如下：

```kotlin
1.  @Component
2.  class BlogTagRelationDAO {
3.      @Autowired
4.      lateinit var queryFactory: JPAQueryFactory
5.      @Autowired
6.      lateinit var blogTagRelationRepository: BlogTagRelationRepository
7.      // 根據 relationId 刪除記錄
8.      @Transactional
9.      fun deleteByPrimaryKey(relationId: Long): Int {
10.         val qBlogTagRelation = QBlogTagRelation.blogTagRelation
11.         return queryFactory.delete(qBlogTagRelation)
12.                 .where(qBlogTagRelation.relationId.eq(relationId))
13.                 .execute().toInt()
14.     }
15.     // 儲存資料
16.     @Transactional
17.     fun insert(record: BlogTagRelation): Int {
18.         blogTagRelationRepository.save(record)
19.         return 0
20.     }
21.     // 儲存資料
22.     @Transactional
23.     fun insertSelective(record: BlogTagRelation): Int {
24.         blogTagRelationRepository.save(record)
25.         return 0
26.     }
27.     // 根據 relationId 尋找
28.     fun selectByPrimaryKey(relationId: Long): BlogTagRelation {
29.         val qBlogTagRelation = QBlogTagRelation.blogTagRelation
30.         return queryFactory.selectFrom(qBlogTagRelation)
```

```
31.                .where(qBlogTagRelation.relationId.eq(relationId))
32.                .fetchFirst()
33.        }
34.    // 根據 blogId、tagId 尋找
35.    fun selectByBlogIdAndTagId(blogId: Long, tagId: Int): BlogTagRelation {
36.        val qBlogTagRelation = QBlogTagRelation.blogTagRelation
37.        return queryFactory.selectFrom(qBlogTagRelation)
38.                .where(qBlogTagRelation.blogId.eq(blogId).and
    (qBlogTagRelation.tagId.eq(tagId)))
39.                .fetchFirst()
40.    }
41.    // 查閱資料表中存在的 tagIds
42.    fun selectDistinctTagIds(tagIds: List<Int>): List<Int> {
43.        val qBlogTagRelation = QBlogTagRelation.blogTagRelation
44.        return queryFactory.selectDistinct(qBlogTagRelation.tagId)
45.                .from(qBlogTagRelation)
46.                .where(qBlogTagRelation.tagId.`in`(tagIds))
47.                .fetchResults()
48.                .results
49.    }
50.    // 將 record 不為 null 的欄位更新到資料庫
51.    @Transactional
52.    fun updateByPrimaryKeySelective(record: BlogTagRelation): Int {
53.        // 省略部分程式……
54.        return queryFactory.update(qBlogTagRelation)
55.                .set(cols, values)
56.                .where(qBlogTagRelation.relationId.eq(record.relationId))
57.                .execute()
58.                .toInt()
59.    }
60.    // 更新 BlogTagRelation
61.    @Transactional
62.    fun updateByPrimaryKey(record: BlogTagRelation): Int {
63.        // 省略部分程式……
64.    }
```

```
65.     // 批次插入 BlogTagRelation
66.     @Transactional
67.     fun batchInsert(blogTagRelationList: List<BlogTagRelation>): Int {
68.         blogTagRelationRepository.saveAll(blogTagRelationList)
69.         return 0
70.     }
71.     // 根據 blogId 刪除
72.     @Transactional
73.     fun deleteByBlogId(blogId: Long?): Int {
74.         val qBlogTagRelation = QBlogTagRelation.blogTagRelation
75.         return queryFactory.delete(qBlogTagRelation)
76.                 .where(qBlogTagRelation.blogId.eq(blogId))
77.                 .execute().toInt()
78.     }
79. }
```

10.6 Service 層的設計

部落格定義的服務層如表 10.4 所示。

表 10.4 部落格定義的服務清單

序號	服務	說明
1	AdminUserService	管理使用者服務
2	BlogService	部落格服務
3	CategoryService	分類服務
4	CommentService	評論服務
5	ConfigService	設定服務
6	LinkService	連結（友情連結）服務
7	TagService	標籤服務

AdminUserService.kt 提供了管理使用者相關服務：登入、取得使用者資訊、修改目前登入使用者的密碼及修改目前登入使用者的名稱資訊。程式如下：

```kotlin
1.  interface AdminUserService {
2.      // 登入
3.      fun login(userName: String, password: String): AdminUser?
4.      /**
5.       * 取得使用者資訊
6.       * @param loginUserId
7.       * @return
8.       */
9.      fun getUserDetailById(loginUserId: Int): AdminUser?
10.     /**
11.      * 修改目前登入使用者的密碼
12.      * @param loginUserId
13.      * @param originalPassword
14.      * @param newPassword
15.      * @return
16.      */
17.     fun updatePassword(loginUserId: Int, originalPassword: String,
    newPassword: String): Boolean
18.     /**
19.      * 修改目前登入使用者的名稱資訊
20.      * @param loginUserId
21.      * @param loginUserName
22.      * @param nickName
23.      * @return
24.      */
25.     fun updateName(loginUserId: Int, loginUserName: String, nickName:
    String): Boolean
26. }
```

BlogService.kt 提供了部落格相關服務：儲存部落格內容、分頁查詢部落格、批次刪除部落格、取得部落格總數、根據 id 取得詳情、更新部落格、取得首頁文章列表、查詢首頁側邊欄顯示的部落格、取得文章詳情、根據分類取得文章清單及根據搜尋框的輸入取得文章列表等。程式如下：

```
1.  interface BlogService {
2.      // 儲存部落格
3.      fun saveBlog(blog: Blog): String
4.      // 分頁查詢部落格
5.      fun getBlogsPage(pageUtil: PageQueryUtil): PageResult
6.      // 批次刪除部落格
7.      fun deleteBatch(ids: List<Long>): Boolean
8.      // 取得部落格總數
9.      fun getTotalBlogs(): Int
10.     /**
11.      * 根據 id 取得詳情
12.      * @param blogId
13.      * @return
14.      */
15.     fun getBlogById(blogId: Long): Blog
16.     /**
17.      * 後台修改
18.      * @param blog
19.      * @return
20.      */
21.     fun updateBlog(blog: Blog): String
22.     /**
23.      * 取得首頁文章列表
24.      * @param page
25.      * @return
26.      */
27.     fun getBlogsForIndexPage(page: Int): PageResult
28.     /**
29.      * 取得首頁側邊欄資料列表
30.      * 0- 點擊最多 1- 最新發佈
31.      * @param type
32.      * @return
```

```
33.     */
34.     fun getBlogListForIndexPage(type: Int): List<SimpleBlogListVO>
35.     /**
36.      * 文章詳情
37.      * @param blogId
38.      * @return
39.      */
40.     fun getBlogDetail(blogId: Long): BlogDetailVO?
41.     /**
42.      * 根據標籤取得文章清單
43.      * @param tagName
44.      * @param page
45.      * @return
46.      */
47.     fun getBlogsPageByTag(tagName: String, page: Int): PageResult?
48.     /**
49.      * 根據分類取得文章清單
50.      * @param categoryId
51.      * @param page
52.      * @return
53.      */
54.     fun getBlogsPageByCategory(categoryId: String, page: Int): PageResult?
55.     /**
56.      * 根據搜尋取得文章清單
57.      * @param keyword
58.      * @param page
59.      * @return
60.      */
61.     fun getBlogsPageBySearch(keyword: String, page: Int): PageResult?
62.     //  根據 subUrl 尋找部落格明細
63.     fun getBlogDetailBySubUrl(subUrl: String): BlogDetailVO?
64. }
```

CategoryService.kt 提供了分類相關服務：分頁查詢分類資訊及增加分類資料等。程式如下：

```
1.   interface CategoryService {
2.       /**
3.        *  查詢分類的分頁資料
4.        *  @param pageUtil
5.        *  @return
6.        */
7.       fun getBlogCategoryPage(pageUtil: PageQueryUtil): PageResult
8.       //  查詢分類總數
9.       fun getTotalCategories(): Int
10.      /**
11.       *  增加分類資料
12.       *  @param categoryName
13.       *  @param categoryIcon
14.       *  @return
15.       */
16.      fun saveCategory(categoryName: String, categoryIcon: String): Boolean
17.      //  更新分類名稱及圖示
18.      fun updateCategory(categoryId: Int, categoryName: String,
     categoryIcon: String): Boolean
19.      //  批次刪除
20.      fun deleteBatch(ids: List<Int>): Boolean
21.      //  尋找所有分類
22.      fun getAllCategories(): List<BlogCategory>
23.  }
```

CommentService.kt 提供了評論相關服務：增加評論、分頁查詢評論、批次審核評論、批次刪除評論、對評論增加回覆、根據文章 id 和分頁參數取得文章的評論列表等。程式如下：

```
1.   interface CommentService {
2.       /**
3.        *  增加評論
4.        *  @param blogComment
5.        *  @return
6.        */
```

```
7.      fun addComment(blogComment: BlogComment): Boolean
8.      /**
9.       * 後台管理系統中的評論分頁功能
10.      * @param pageUtil
11.      * @return
12.      */
13.     fun getCommentsPage(pageUtil: PageQueryUtil): PageResult
14.     // 查詢評論總數
15.     fun getTotalComments(): Int
16.     /**
17.      * 批次審核
18.      * @param ids
19.      * @return
20.      */
21.     fun checkDone(ids: List<Long>): Boolean
22.     /**
23.      * 批次刪除
24.      * @param ids
25.      * @return
26.      */
27.     fun deleteBatch(ids: List<Long>): Boolean
28.     /**
29.      * 增加回覆
30.      * @param commentId
31.      * @param replyBody
32.      * @return
33.      */
34.     fun reply(commentId: Long, replyBody: String): Boolean
35.     /**
36.      * 根據文章 id 和分頁參數取得文章的評論列表
37.      * @param blogId
38.      * @param page
39.      * @return
40.      */
41.     fun getCommentPageByBlogIdAndPageNum(blogId: Long, page: Int):
```

```
     PageResult?
42. }
```

ConfigService.kt 提供了設定相關服務：修改設定項目及取得所有的設定項目等。程式如下：

```
1.   interface ConfigService {
2.       /**
3.        * 修改設定項目
4.        * @param configName
5.        * @param configValue
6.        * @return
7.        */
8.       fun updateConfig(configName: String, configValue: String): Int
9.       /**
10.       * 取得所有的設定項目
11.       * @return
12.       */
13.      fun getAllConfigs(): Map<String, String>
14. }
```

LinkService.kt 提供了友情連結相關服務：分頁查詢友情連結、查詢友情連結總數、儲存友情連結、根據 id 查詢友情連結、更新友情連結內容、批次刪除友情連結、傳回友情連結頁面所需的所有資料等。程式如下：

```
1.   interface LinkService {
2.       /**
3.        * 查詢友情連結的分頁資料
4.        * @param pageUtil
5.        * @return
6.        */
7.       fun getBlogLinkPage(pageUtil: PageQueryUtil): PageResult
8.       // 取得友情連結總數
9.       fun getTotalLinks(): Int
10.      // 儲存友情連結
```

```
11.     fun saveLink(link: BlogLink): Boolean
12.     //   根據 id 查詢友情連結
13.     fun selectById(id: Int?): BlogLink
14.     //   更新友情連結
15.     fun updateLink(tempLink: BlogLink): Boolean
16.     //   批次刪除
17.     fun deleteBatch(ids: List<Int>): Boolean
18.     /**
19.      * 傳回友情連結頁面所需的所有資料
20.      * @return
21.      */
22.     fun getLinksForLinkPage(): Map<Byte, List<BlogLink>>
23. }
```

TagService.kt 提供了標籤服務：分頁查詢標籤資料、取得標籤總數、儲存標籤、批次刪除標籤及取得每個標籤對應的部落格總數等。程式如下：

```
1.  interface TagService {
2.      /**
3.       * 查詢標籤的分頁資料
4.       * @param pageUtil
5.       * @return
6.       */
7.      fun getBlogTagPage(pageUtil: PageQueryUtil): PageResult
8.      //   查詢標籤總數
9.      fun getTotalTags(): Int
10.     //   儲存標籤
11.     fun saveTag(tagName: String): Boolean
12.     //   批次刪除
13.     fun deleteBatch(ids: List<Int>): Boolean
14.     //   取得每個標籤對應的部落格總數
15.     fun getBlogTagCountForIndex(): List<BlogTagCount>
16. }
```

10.7 Controller 層的設計

部落格定義的介面層如表 10.5 所示。

表 10.5 部落格定義的介面清單

序號	controller	說明
前台介面		
1	MyBlogController	提供部落格前台展示相關介面
後台介面		
2	AdminController	後台管理相關介面
3	BlogController	部落格管理相關介面
4	CategoryController	類別管理相關介面
5	CommentController	評論管理相關介面
6	ConfigurationController	設定管理相關介面
7	LinkController	連結管理相關介面
8	TagController	標籤管理相關介面
9	UploadController	上傳相關介面

AdminController.kt 提供了後台管理相關介面和視圖跳躍：登入、跳躍到管理後台首頁、取得使用者資訊、修改密碼、修改使用者名稱及登出等。程式如下：

```
1.  @Controller
2.  @RequestMapping("/admin")
3.  class AdminController {
4.      @Resource
5.      lateinit var adminUserService: AdminUserService
6.      @Resource
7.      lateinit var blogService: BlogService
```

```kotlin
8.      @Resource
9.      lateinit var categoryService: CategoryService
10.     @Resource
11.     lateinit var linkService: LinkService
12.     @Resource
13.     lateinit var tagService: TagService
14.     @Resource
15.     lateinit var commentService: CommentService
16.     // 跳躍到登入頁
17.     @GetMapping("/login")
18.     fun login(): String {
19.         return "admin/login"
20.     }
21.     // 跳躍到管理後台首頁
22.     @GetMapping("", "/", "/index", "/index.html")
23.     fun index(request: HttpServletRequest): String {
24.         // 省略部分程式……
25.         return "admin/index"
26.     }
27.     // 登入
28.     @PostMapping(value = ["/login"])
29.     fun login(@RequestParam("userName") userName: String,
30.              @RequestParam("password") password: String,
31.              @RequestParam("verifyCode") verifyCode: String,
32.              session: HttpSession): String {
33.         if (StringUtils.isEmpty(verifyCode)) {
34.             session.setAttribute("errorMsg", " 驗證碼不能為空 ")
35.             return "admin/login"
36.         }
37.         if (StringUtils.isEmpty(userName) || StringUtils.isEmpty
    (password)) {
38.             session.setAttribute("errorMsg", " 使用者名稱或密碼不能為空 ")
39.             return "admin/login"
40.         }
41.         val kaptchaCode = session?.getAttribute("verifyCode").toString() + ""
```

```
42.         if (StringUtils.isEmpty(kaptchaCode) || verifyCode != kaptchaCode) {
43.             session.setAttribute("errorMsg", " 驗證碼錯誤 ")
44.             return "admin/login"
45.         }
46.         val adminUser = adminUserService.login(userName, password)
47.         if (adminUser != null) {
48.             session.setAttribute("loginUser", adminUser.nickName)
49.             session.setAttribute("loginUserId", adminUser.adminUserId)
50.             // 將 session 過期時間設定為 7200 秒，即兩小時
51.             //session.setMaxInactiveInterval(60 * 60 * 2);
52.             return "redirect:/admin/index"
53.         } else {
54.             session.setAttribute("errorMsg", " 登入失敗 ")
55.             return "admin/login"
56.         }
57.     }
58.     // 跳躍到個人資訊視圖頁
59.     @GetMapping("/profile")
60.     fun profile(request: HttpServletRequest): String {
61.         // 省略部分程式……
62.         return "admin/profile"
63.     }
64.     // 更新使用者密碼
65.     @PostMapping("/profile/password")
66.     @ResponseBody
67.     fun passwordUpdate(request: HttpServletRequest, @RequestParam
    ("originalPassword") originalPassword: String,
68.                        @RequestParam("newPassword") newPassword: String):
    String {
69.         // 省略部分程式……
70.     }
71.     // 更新使用者名稱
72.     @PostMapping("/profile/name")
73.     @ResponseBody
74.     fun nameUpdate(request: HttpServletRequest, @RequestParam
```

```
   ("loginUserName") loginUserName: String,
75.                @RequestParam("nickName") nickName: String): String {
76.         // 省略部分程式……
77.     }
78.     // 跳躍到登出視圖頁
79.     @GetMapping("/logout")
80.     fun logout(request: HttpServletRequest): String {
81.         // 省略部分程式……
82.         return "admin/login"
83.     }
84. }
```

BlogController.kt 提供了部落格管理相關介面和視圖跳躍：展示部落格清單、跳躍到部落格管理頁面、編輯部落格、儲存部落格、更新部落格、上傳部落格封面圖片及刪除部落格等。程式如下：

```
1.  @Controller
2.  @RequestMapping("/admin")
3.  class BlogController {
4.      @Resource
5.      lateinit var blogService: BlogService
6.      @Resource
7.      lateinit var categoryService: CategoryService
8.      // 取得部落格列表
9.      @GetMapping("/blogs/list")
10.     @ResponseBody
11.     fun list(@RequestParam params: Map<String, Any>): Result<Any> {
12.         // 省略部分程式……
13.     }
14.     // 跳躍到部落格視圖頁
15.     @GetMapping("/blogs")
16.     fun list(request: HttpServletRequest): String {
17.         request.setAttribute("path", "blogs")
18.         return "admin/blog"
19.     }
```

```
20.      // 跳躍到部落格編輯視圖頁
21.      @GetMapping("/blogs/edit")
22.      fun edit(request: HttpServletRequest): String {
23.          // 省略部分程式……
24.          return "admin/edit"
25.      }
26.      // 跳躍到部落格編輯視圖頁
27.      @GetMapping("/blogs/edit/{blogId}")
28.      fun edit(request: HttpServletRequest, @PathVariable("blogId") blogId:
     Long): String {
29.          // 省略部分程式……
30.          return "admin/edit"
31.      }
32.      // 儲存部落格
33.      @PostMapping("/blogs/save")
34.      @ResponseBody
35.      fun save(@RequestParam("blogTitle") blogTitle: String,
36.               @RequestParam(name = "blogSubUrl", required = false)
     blogSubUrl: String,
37.               @RequestParam("blogCategoryId") blogCategoryId: Int,
38.               @RequestParam("blogTags") blogTags: String,
39.               @RequestParam("blogContent") blogContent: String,
40.               @RequestParam("blogCoverImage") blogCoverImage: String,
41.               @RequestParam("blogStatus") blogStatus: Byte,
42.               @RequestParam("enableComment") enableComment: Byte):
     Result<Any> {
43.          // 省略部分程式……
44.          val saveBlogResult = blogService.saveBlog(blog)
45.          return if ("success" == saveBlogResult) {
46.              ResultGenerator.genSuccessResult(" 增加成功 ")
47.          } else {
48.              ResultGenerator.genFailResult(saveBlogResult)
49.          }
50.      }
51.      // 更新部落格
```

```kotlin
52.     @PostMapping("/blogs/update")
53.     @ResponseBody
54.     fun update(@RequestParam("blogId") blogId: Long,
55.                 @RequestParam("blogTitle") blogTitle: String,
56.                 @RequestParam(name = "blogSubUrl", required = false)
    blogSubUrl: String,
57.                 @RequestParam("blogCategoryId") blogCategoryId: Int,
58.                 @RequestParam("blogTags") blogTags: String,
59.                 @RequestParam("blogContent") blogContent: String,
60.                 @RequestParam("blogCoverImage") blogCoverImage: String,
61.                 @RequestParam("blogStatus") blogStatus: Byte,
62.                 @RequestParam("enableComment") enableComment: Byte):
    Result<Any> {
63.         // 省略部分程式……
64.         val updateBlogResult = blogService.updateBlog(blog)
65.         return if ("success" == updateBlogResult) {
66.             ResultGenerator.genSuccessResult(" 修改成功 ")
67.         } else {
68.             ResultGenerator.genFailResult(updateBlogResult)
69.         }
70.     }
71.     // 上傳圖片
72.     @PostMapping("/blogs/md/uploadfile")
73.     @Throws(IOException::class, URISyntaxException::class)
74.     fun uploadFileByEditormd(request: HttpServletRequest,
75.                               response: HttpServletResponse,
76.                               @RequestParam(name = "editormd-image-file",
    required = true)
77.                               file: MultipartFile) {
78.         // 省略部分程式……
79.     }
80.     // 批次刪除部落格
81.     @PostMapping("/blogs/delete")
82.     @ResponseBody
83.     fun delete(@RequestBody ids: List<Long>): Result<Any> {
```

```
84.          if (ids.size < 1) {
85.              return ResultGenerator.genFailResult(" 參數異常！")
86.          }
87.          return if (blogService.deleteBatch(ids)) {
88.              ResultGenerator.genSuccessResult()
89.          } else {
90.              ResultGenerator.genFailResult(" 刪除失敗 ")
91.          }
92.      }
93. }
```

CategoryController.kt 提供類別管理介面和視圖跳躍：取得分類列表、增加分類、修改分類、刪除分類、跳躍到分類管理頁面等。程式如下：

```
1.  @Controller
2.  @RequestMapping("/admin")
3.  class CategoryController {
4.      @Resource
5.      lateinit var categoryService: CategoryService
6.      // 跳躍到類別視圖頁
7.      @GetMapping("/categories")
8.      fun categoryPage(request: HttpServletRequest): String {
9.          request.setAttribute("path", "categories")
10.         return "admin/category"
11.     }
12.     /**
13.      * 分類列表
14.      */
15.     @RequestMapping(value = ["/categories/list"], method = [RequestMethod.
    GET])
16.     @ResponseBody
17.     fun list(@RequestParam params: Map<String, Any>): Result<Any> {
18.         if (StringUtils.isEmpty(params["page"]) || StringUtils.
    isEmpty(params["limit"])) {
19.             return ResultGenerator.genFailResult(" 參數異常！")
```

```
20.         }
21.         val pageUtil = PageQueryUtil(params)
22.         return ResultGenerator.genSuccessResult (categoryService!!.
    getBlogCategoryPage(pageUtil))
23.     }
24.     /**
25.      * 增加分類
26.      */
27.     @RequestMapping(value = ["/categories/save"], method = [RequestMethod.
    POST])
28.     @ResponseBody
29.     fun save(@RequestParam("categoryName") categoryName: String,
30.             @RequestParam("categoryIcon") categoryIcon: String):
    Result<Any> {
31.         // 省略部分程式……
32.     }
33.     /**
34.      * 修改分類
35.      */
36.     @RequestMapping(value = ["/categories/update"], method =
    [RequestMethod.POST])
37.     @ResponseBody
38.     fun update(@RequestParam("categoryId") categoryId: Int,
39.             @RequestParam("categoryName") categoryName: String,
40.             @RequestParam("categoryIcon") categoryIcon: String):
    Result<Any> {
41.         // 省略部分程式……
42.     }
43.     /**
44.      * 刪除分類
45.      */
46.     @RequestMapping(value = ["/categories/delete"], method =
    [RequestMethod.POST])
47.     @ResponseBody
48.     fun delete(@RequestBody ids: List<Int>): Result<Any> {
```

```
49.        // 省略部分程式……
50.    }
51. }
```

CommentController.kt 提供了評論管理介面和視圖跳躍：取得評論列表、審核評論、回覆評論、刪除評論、跳躍到評論管理頁面等。程式如下：

```
1.  @Controller
2.  @RequestMapping("/admin")
3.  class CommentController {
4.      @Resource
5.      lateinit var commentService: CommentService
6.      // 取得評論列表
7.      @GetMapping("/comments/list")
8.      @ResponseBody
9.      fun list(@RequestParam params: Map<String, Any>): Result<Any> {
10.         // 省略部分程式……
11.     }
12.     // 審核評論
13.     @PostMapping("/comments/checkDone")
14.     @ResponseBody
15.     fun checkDone(@RequestBody ids: List<Long>): Result<Any> {
16.         // 省略部分程式……
17.     }
18.     // 回覆評論
19.     @PostMapping("/comments/reply")
20.     @ResponseBody
21.     fun checkDone(@RequestParam("commentId") commentId: Long?,
22.                   @RequestParam("replyBody") replyBody: String):
    Result<Any> {
23.         // 省略部分程式……
24.  }
25.     // 刪除評論
26.     @PostMapping("/comments/delete")
27.     @ResponseBody
```

```
28.    fun delete(@RequestBody ids: List<Long>): Result<Any> {
29.        // 省略部分程式……
30.    }
31.    // 跳躍到評論視圖頁
32.    @GetMapping("/comments")
33.    fun list(request: HttpServletRequest): String {
34.        request.setAttribute("path", "comments")
35.        return "admin/comment"
36.    }
37. }
```

ConfigurationController.kt 提供了設定管理介面和視圖跳躍：跳躍到設定管理頁面、更新網站相關設定、更新使用者相關設定及更新網站頁尾相關設定等。程式如下：

```
1.  @Controller
2.  @RequestMapping("/admin")
3.  class ConfigurationController {
4.      @Resource
5.      lateinit var configService: ConfigService
6.      // 跳躍到設定視圖頁
7.      @GetMapping("/configurations")
8.      fun list(request: HttpServletRequest): String {
9.          // 省略部分程式……
10.         return "admin/configuration"
11.     }
12.     // 取得部落格 website 設定
13.     @PostMapping("/configurations/website")
14.     @ResponseBody
15.     fun website(@RequestParam(value = "websiteName", required = false)
    websiteName: String,
16.                 @RequestParam(value = "websiteDescription", required =
    false) websiteDescription: String,
17.                 @RequestParam(value = "websiteLogo", required = false)
    websiteLogo: String,
```

```
18.                 @RequestParam(value = "websiteIcon", required = false)
      websiteIcon: String): Result<Any> {
19.          // 省略部分程式……
20.      }
21.      // 取得部落格 userInfo 設定
22.      @PostMapping("/configurations/userInfo")
23.      @ResponseBody
24.      fun userInfo(@RequestParam(value = "yourAvatar", required = false)
      yourAvatar: String,
25.                  @RequestParam(value = "yourName", required = false)
      yourName: String,
26.                  @RequestParam(value = "yourEmail", required = false)
      yourEmail: String): Result<Any> {
27.          // 省略部分程式……
28.      }
29.      // 取得部落格 footer 設定
30.      @PostMapping("/configurations/footer")
31.      @ResponseBody
32.      fun footer(@RequestParam(value = "footerAbout", required = false)
      footerAbout: String,
33.                  @RequestParam(value = "footerICP", required = false)
      footerICP: String,
34.                  @RequestParam(value = "footerCopyRight", required = false)
      footerCopyRight: String,
35.                  @RequestParam(value = "footerPoweredBy", required = false)
      footerPoweredBy: String,
36.                  @RequestParam(value = "footerPoweredByURL", required =
      false) footerPoweredByURL: String): Result<Any> {
37.          // 省略部分程式…
38.      }
39. }
```

LinkController.kt 提供連結（友情連結）管理相關介面和視圖跳躍：跳躍到
友情連結管理頁面、取得友情連結清單、增加友情連結、取得友情連結詳
情、修改友情連結及刪除友情連結等。程式如下：

```
1.  @Controller
2.  @RequestMapping("/admin")
3.  class LinkController {
4.      @Resource
5.      lateinit var linkService: LinkService
6.      // 跳躍到友情連結視圖
7.      @GetMapping("/links")
8.      fun linkPage(request: HttpServletRequest): String {
9.          request.setAttribute("path", "links")
10.         return "admin/link"
11.     }
12.     // 取得友情連結列表
13.     @GetMapping("/links/list")
14.     @ResponseBody
15.     fun list(@RequestParam params: Map<String, Any>): Result<Any> {
16.         // 省略部分程式……
17.     }
18.     /**
19.      * 增加友情連結
20.      */
21.     @RequestMapping(value = ["/links/save"], method = [RequestMethod.POST])
22.     @ResponseBody
23.     fun save(@RequestParam("linkType") linkType: Int?,
24.             @RequestParam("linkName") linkName: String,
25.             @RequestParam("linkUrl") linkUrl: String,
26.             @RequestParam("linkRank") linkRank: Int?,
27.             @RequestParam("linkDescription") linkDescription: String):
    Result<Any> {
28.         // 省略部分程式……
29.     }
30.     /**
31.      * 詳情
32.      */
33.     @GetMapping("/links/info/{id}")
34.     @ResponseBody
```

```
35.        fun info(@PathVariable("id") id: Int?): Result<Any> {
36.            // 省略部分程式……
37.        }
38.        /**
39.         * 修改友情連結
40.         */
41.        @RequestMapping(value = ["/links/update"], method = [RequestMethod.POST])
42.        @ResponseBody
43.        fun update(@RequestParam("linkId") linkId: Int?,
44.                   @RequestParam("linkType") linkType: Int?,
45.                   @RequestParam("linkName") linkName: String,
46.                   @RequestParam("linkUrl") linkUrl: String,
47.                   @RequestParam("linkRank") linkRank: Int?,
48.                   @RequestParam("linkDescription") linkDescription: String):
       Result<Any> {
49.            // 省略部分程式……
50.        }
51.        /**
52.         * 刪除友情連結
53.         */
54.        @RequestMapping(value = ["/links/delete"], method = [RequestMethod.POST])
55.        @ResponseBody
56.        fun delete(@RequestBody ids: List<Int>): Result<Any> {
57.            // 省略部分程式……
58.        }
59. }
```

TagController.kt 提供了標籤管理介面和視圖跳躍：跳躍到標籤管理頁面、取得標籤清單、儲存標籤及刪除標籤等。程式如下：

```
1.  @Controller
2.  @RequestMapping("/admin")
3.  class TagController {
4.      @Resource
5.      lateinit var tagService: TagService
```

```
6.        // 跳躍到標籤視圖
7.        @GetMapping("/tags")
8.        fun tagPage(request: HttpServletRequest): String {
9.            request.setAttribute("path", "tags")
10.            return "admin/tag"
11.        }
12.        // 取得標籤列表
13.        @GetMapping("/tags/list")
14.        @ResponseBody
15.        fun list(@RequestParam params: Map<String, Any>): Result<Any> {
16.            // 省略部分程式……
17.        }
18.        // 儲存標籤
19.        @PostMapping("/tags/save")
20.        @ResponseBody
21.        fun save(@RequestParam("tagName") tagName: String): Result<Any> {
22.            // 省略部分程式……
23.        }
24.        // 刪除標籤
25.        @PostMapping("/tags/delete")
26.        @ResponseBody
27.        fun delete(@RequestBody ids: List<Int>): Result<Any> {
28.            // 省略部分程式……
29.        }
30. }
```

UploadController.kt 提供了上傳檔案的介面：

```
1.  @Controller
2.  @RequestMapping("/admin")
3.  class UploadController {
4.        // 上傳檔案
5.        @PostMapping("/upload/file")
6.        @ResponseBody
7.        @Throws(URISyntaxException::class)
8.        fun upload(httpServletRequest: HttpServletRequest, @RequestParam
```

```
       ("file") file: MultipartFile): Result<Any> {
9.           // 省略部分程式……
10.        }
11. }
```

MyBlogController.kt 提供了前台展示相關介面和視圖跳躍：跳躍到部落格首頁、跳躍到分類頁面、跳躍到部落格詳情頁面、跳躍到標籤清單頁面、跳躍到搜尋清單頁、跳躍到友情連結頁、增加文章評論等。程式如下：

```
1.  @Controller
2.  class MyBlogController {
3.      @Resource
4.      lateinit var blogService: BlogService
5.      @Resource
6.      lateinit var tagService: TagService
7.      @Resource
8.      lateinit var linkService: LinkService
9.      @Resource
10.     lateinit var commentService: CommentService
11.     @Resource
12.     lateinit var configService: ConfigService
13.     @Resource
14.     lateinit var categoryService: CategoryService
15.     /**
16.      * 首頁
17.      * @return
18.      */
19.     @GetMapping("/", "/index", "index.html")
20.     fun index(request: HttpServletRequest): String {
21.         return this.page(request, 1)
22.     }
23.     /**
24.      * 首頁 分頁資料
25.      * @return
26.      */
```

```kotlin
27.     @GetMapping("/page/{pageNum}")
28.     fun page(request: HttpServletRequest, @PathVariable("pageNum")
    pageNum: Int): String {
29.         // 省略部分程式……
30.         return "blog/$theme/index"
31.     }
32.     /**
33.      * Categories 頁面 ( 包含分類資料和標籤資料 )
34.      * @return
35.      */
36.     @GetMapping("/categories")
37.     fun categories(request: HttpServletRequest): String {
38.         // 省略部分程式……
39.         return "blog/$theme/category"
40.     }
41.     /**
42.      * 詳情頁
43.      * @return
44.      */
45.     @GetMapping("/blog/{blogId}", "/article/{blogId}")
46.     fun detail(request: HttpServletRequest, @PathVariable("blogId")
    blogId: Long, @RequestParam(value = "commentPage", required = false,
    defaultValue = "1") commentPage: Int): String {
47.         // 省略部分程式……
48.         return "blog/$theme/detail"
49.     }
50.     /**
51.      * 標籤列表頁
52.      * @return
53.      */
54.     @GetMapping("/tag/{tagName}")
55.     fun tag(request: HttpServletRequest, @PathVariable("tagName") tagName:
    String): String {
56.         return tag(request, tagName, 1)
57.     }
```

```
58.    /**
59.     * 標籤列表頁
60.     * @return
61.     */
62.    @GetMapping("/tag/{tagName}/{page}")
63.    fun tag(request: HttpServletRequest, @PathVariable("tagName") tagName:
   String, @PathVariable("page") page: Int): String {
64.        // 省略部分程式……
65.        return "blog/$theme/list"
66.    }
67.    /**
68.     * 分類列表頁
69.     * @return
70.     */
71.    @GetMapping("/category/{categoryName}")
72.    fun category(request: HttpServletRequest, @PathVariable
   ("categoryName") categoryName: String): String {
73.        return category(request, categoryName, 1)
74.    }
75.    /**
76.     * 分類列表頁
77.     * @return
78.     */
79.    @GetMapping("/category/{categoryName}/{page}")
80.    fun category(request: HttpServletRequest, @PathVariable("categoryName")
   categoryName: String, @PathVariable("page") page: Int): String {
81.        // 省略部分程式……
82.        return "blog/$theme/list"
83.    }
84.    /**
85.     * 搜尋列表頁
86.     * @return
87.     */
88.    @GetMapping("/search/{keyword}")
```

```kotlin
89.    fun search(request: HttpServletRequest, @PathVariable("keyword")
   keyword: String): String {
90.        return search(request, keyword, 1)
91.    }
92.    /**
93.     * 搜尋列表頁
94.     * @return
95.     */
96.    @GetMapping("/search/{keyword}/{page}")
97.    fun search(request: HttpServletRequest, @PathVariable("keyword")
   keyword: String, @PathVariable("page") page: Int): String {
98.        // 省略部分程式……
99.        return "blog/$theme/list"
100.        }
101.    /**
102.     * 友情連結頁
103.     * @return
104.     */
105.        @GetMapping("/link")
106.        fun link(request: HttpServletRequest): String {
107.            // 省略部分程式……
108.            return "blog/$theme/link"
109.        }
110.    /**
111.     * 評論操作
112.     */
113.        @PostMapping(value = ["/blog/comment"])
114.        @ResponseBody
115.        fun comment(request: HttpServletRequest, session: HttpSession,
116.                    @RequestParam blogId: Long?, @RequestParam verifyCode:
   String,
117.                    @RequestParam commentator: String, @RequestParam
   email: String,
118.                    @RequestParam websiteUrl: String, @RequestParam
```

```
        commentBody: String): Result<Any> {
119.            // 省略部分程式……
120.            return ResultGenerator.genSuccessResult(commentService.
     addComment(comment))
121.        }
122.        /**
123.         * 關於頁面以及其他設定了 subUrl 的文章頁
124.         * @return
125.         */
126.        @GetMapping("/{subUrl}")
127.        fun detail(request: HttpServletRequest, @PathVariable("subUrl")
     subUrl: String): String {
128.            // 省略部分程式……
129.        }
130.        companion object {
131.            //public static String theme = "default";
132.            //public static String theme = "yummy-jekyll";
133.            var theme = "amaze"
134.        }
135.    }
```

10.8 部署到騰訊雲

在騰訊雲上安裝 JDK 1.8 和 MySQL。將子專案 chapter11-blog 用 Maven 包
裝成一個 Jar 檔案,透過 xFTP 工具上傳到騰訊雲主機,並上傳設定檔,修
改資料庫連接設定,應用目錄如圖 10.3 所示。在 MySQL 資料庫中建表,
並初始化資料。執行以下指令,啟動部落格服務,如圖 10.4 所示。

```
nohup java -jar chapter11-blog-0.0.1-SNAPSHOT.jar > blog.log &
```

```
drwxrwxr-x 2 ubuntu ubuntu     4096 Apr 23 21:53 ./
drwxrwxr-x 3 ubuntu ubuntu     4096 Apr 23 21:24 ../
-rw-rw-r-- 1 ubuntu ubuntu      556 Apr 23 21:42 application-prd.yml
-rw-rw-r-- 1 ubuntu ubuntu      175 Apr 23 21:26 application.yml
-rw-rw-r-- 1 ubuntu ubuntu    59181 May 26 13:35 blog.log
-rw-rw-r-- 1 ubuntu ubuntu 68442906 Apr 23 21:25 chapter11-blog-0.0.1-SNAPSHOT.jar
```

圖 10.3　騰訊雲端應用目錄

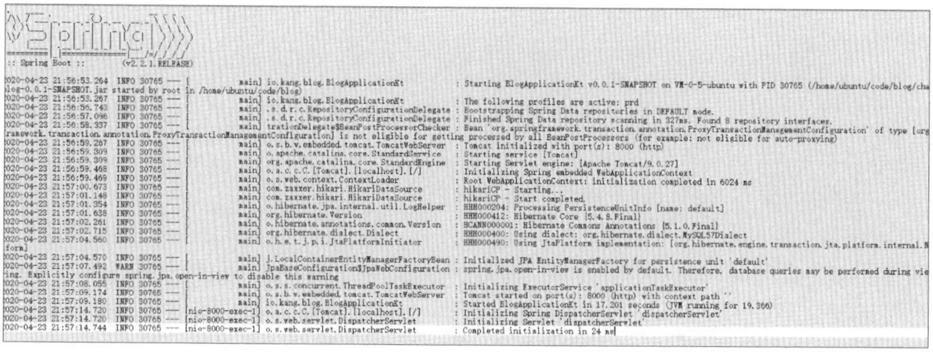

圖 10.4　部落格開機記錄

10.9 小結

本章透過一個部落格實例介紹了如何使用 Kotlin、Spring Boot 開發一個簡單的應用。本應用採用分層結構，自下而上分為資料物理層、資料庫操作層、應用服務層及介面層。本章透過大量原始程式詳細介紹了各層的開發方法，最後介紹了如何將部落格應用部署到騰訊雲，並展示了部落格的效果。